Site Selection: Finding and Developing Your Best Location

Kay Whitehouse, CCIM

LIBERTY HOUSE®

Dedicated to my mother, Jessie Lewis Vance,
who always told me I could do anything I set my mind to,
and to my two daughters,
Theda Lynn Whitehouse and Valerie Kay Whitehouse,
my two constant sources of inspiration.

First Edition • First Printing

Library of Congress Cataloging-in-Publication Data

Whitehouse, Kay.
Site selection : finding and developing your best location / by Kay Whitehouse.
p. cm.
Includes index.
ISBN 0-8306-3053-8
1. Business enterprises—Location. 2. Real estate development.
I. Title.
HD58.W38 1987
658.2'1—dc20 89-8294
CIP

TAB BOOKS Inc. offers software for sale. For information and a catalog please contact TAB Software Department, Blue Ridge Summit, PA 17294-0850.

Questions regarding the content of this book should be addressed to:

Reader Inquiry Branch
TAB BOOKS Inc.
Blue Ridge Summit, PA 17294-0214

Acquisitions Editor: Kimberly Tabor
Technical Editor: Lori Flaherty
Production: Katherine Brown

Contents

Acknowledgments

A special thanks to Kathy Holley at Ideas Unlimited. All photos and illustrations were created by Kathy, whose unique talents were a godsend to me.

Thanks also to my friends who have been so supportive and to God for giving me the whole idea.

Preface

Site Selection: Finding and Developing Your Best Location was written to provide insight into site selection and some of the problems connected with real estate development. Information for any kind of site you need, in your hometown or anywhere in the United States, is included.

This book is about finding the area, the best site in that area, and identifying and dealing with the competition. The pictures and illustrations are included as examples and memory tools.

Selecting a site in a tourist area and moving your company headquarters are outlined more specifically because of their special requirements. Once you determine where you want to be with your tourist project or headquarters, the next step is to return to the basic ABC's: Area, Best Site, and Competition.

Careful planning of a branch office or choosing a second location is also included. These important topics are covered in depth, in order to help you achieve success.

A brief look at the feasibility study can help you determine what you really want to know after all is said and done—will I make any money, when, and how much?

A very popular chapter ''Fifty Flags to Site Selection,'' are red flags to watch out for when selecting and developing real estate. I have run into every one of these situations myself, and I hope that by making you aware of them, you can save time, money—and disappointment.

And what is all this work worth without the decision to move forward. Careful decision making as well as selecting the right professionals to help you can give you the big end result you want: cash flow.

Whether you are a college student, investor, or developer, *Site Selection: Finding and Developing Your Best Location* should be "required reading." This book includes all of the basic information on commercial real estate development and can be used as a training tool for corporate real estate companies.

No other book provides this basic, easy to read and understandable information—including a 170-word glossary of definitions.

Introduction

In traveling around the Sunbelt and talking to people from all walks of life who were trying to buy real estate for one reason or another, it appeared that these people all had something in common—the need to know where to start.

This book is meant as a basic guide in learning where to start when you want to buy real estate, need a site for a project, or for your own personal use.

Where you go with this information is up to you. You might build a shopping center, develop a campground to operate when you retire, or learn where to go to check on future planned development for the section of town you are considering for your home.

Real estate is exciting and fun, and whatever your project or wherever this book might lead you, I wish you good luck.

SECTION 1

1

Getting Started

What would you do if a client of yours, or your company, asked you to go to a town you did not know and find a piece of property suitable for a commercial building? Perhaps it would go something like this:

You have only a limited amount of time, so you hop off the plane, grab a map, rent a car, and drive from one end of town to the other looking for a piece of property that might work. You are exhausted, you cannot remember one area of town from another, and you have wasted a whole day.

Well, it doesn't have to be this way. No matter what the use for your site is, you will find that you have ample information for more than one user because you are going to learn quite a bit about the city and area. A user is a client of yours, or the company you work for (or yourself), who will buy, lease or trade for this piece of property or site and use it for a specific purpose, such as a bank, day care center, restaurant, shopping center, office building, as an investment, or even to land bank for future development.

I know when you get to your city or area, your first impulse is to drive around a while and get acclimated or acquainted. Well, if you have time to waste and want to do that, it's up to you. But, when you are ready to go to work. . . let's go.

THE MAP STORE

First things first. Get a copy of the yellow pages of the telephone book and look under maps. You can do this from the airport, but it would probably be better to go to your motel and check in first. Remember, we are being very organized here. You can get a little map at the rental car counter to get you started.

All checked into your room? Now, look in the yellow pages under maps and try to locate a map store large enough to have a selection of area maps, other facts about the area and information on how the town relates to the rest of the state.

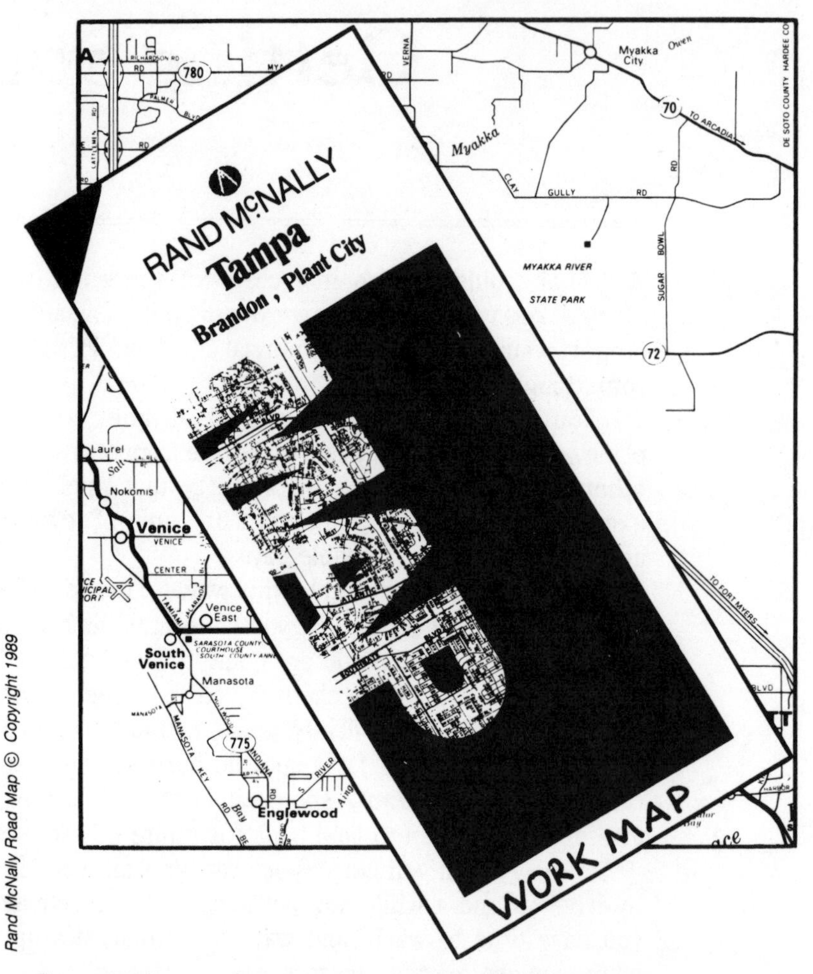

Get enough maps for your presentation and for a work map.

When you go to the map store, and you might want to go to more than one, be sure to spend some time and see what they have that can be of help to you. Open the maps and look them over carefully. Check one against the other. You might be able to read one better than the other or you might prefer the colors or road definition of one to the other.

Be sure that the city limits are defined, and that shopping centers, schools, colleges, hospitals, airports, large offices, parks, industrial plants, subdivisions, tourist attractions or historic sites, the downtown area, interstate or expressways, lakes, rivers or coastline, or anything else you might think of that would be of help to you, is included on the map. All of this information helps you to get to know the town.

Get enough maps so that you can have a work map as well as extras for your presentation to your boss or client. Start with your work map and mark it work map. Clearly mark all of the places I mentioned previously, such as shopping centers, schools, subdivisions, etc.

THE CHAMBER OF COMMERCE

Once you have finished visiting the map stores, and gathering the information you need, it's time to make a stop at the local chamber of commerce.

You will want to speak to the director of the chamber of commerce. The chamber of commerce director can give you invaluable insight to the personality of the city and county; if it is growing (if so, at what rate), where it is growing and what growth is wanted.

Unless your project is completely confidential, tell him why you are interested in the area. See how he reacts to your plans.

All of this information is very important to you and you will want to discuss all of these topics with the director. If the director is busy, it would be best to either wait until he is free, or make an appointment and come back.

Most chamber of commerce offices have brochures, fact sheets, and pamphlets about the city and county (see the sample shown in Appendix A). These publications should show the demographic profile of the area including industries, businesses and other employment generators. The cost of taxes and utilities should also be outlined as well as schools, parks, and other attractions. The cost of living, number of hotels, restaurants, vehicles registered, residential and commercial building permits issued, and much more useful information will be available through the chamber of commerce. Even the average temperature and life style of residents can be provided by the chamber of commerce.

In some areas, particularly large metropolitan areas, the economic development commission would have this information. In this instance, the economic development commission provides this information and can be very instrumental to your project, particularly larger projects.

Information on the area is necessary for you to determine if your project will work in this city or county.

Ask lots of questions, in fact, ask the director of the chamber of commerce or economic development commission if you have missed anything.

CHOOSING A COMMERCIAL REAL ESTATE BROKER

In my opinion, there is only one way to pick a real estate broker; choose one who has the CCIM designation. The CCIM designation is given only to candidates who have successfully completed an intense graduate-level education, practical applications, and who are well seasoned in the commercial real estate field.

The program was designed by the Commercial Investment Real Estate Council of the Realtors National Marketing Institute in Chicago, Illinois.

The CCIM designation signifies that this person is a Certified Commercial Investment Member of the Realtors National Marketing Institute. I could go on and on about this program, and the professionalism of the people involved, but after you have spoken to a real estate broker who has been awarded this prestigious designation, you will understand.

Interview at least three brokers and pick the one you feel knows the market and understands your needs best. One broker's personality might suit you better, or might have a specialization that you feel can help. For example, if your project is developing a retirement community, one of the brokers you interview might have handled a similar project previously; or perhaps handles retirement communities exclusively throughout the state or nation.

Once you have chosen a real estate broker, you are ready to head for the county courthouse and city hall.

2

The Courthouse and City Hall

When you are ready for the courthouse, dress comfortably, you are there to work. Save your, "I'm going to impress you" business suit for your presentation to your client or boss.

Be sure to allow plenty of time. You won't be able to get all the information you need if you arrive at 4:30 p.m. and 5:00 is quitting time.

If you plan on having a site within the city limits, you will eventually want to go to city hall. It is good, however, to start at the county courthouse to get a broader picture of the county before proceeding to city hall.

Once you get to the courthouse, there are several departments you will want to visit. In fact, you are going to want to go into every department in both the county courthouse, and city hall, that has anything to do with real estate and development.

Let's start with the planning and zoning department, mainly because this is where it all starts. The planning and zoning departments have something that can mean life or death to your project and that is: The Comprehensive Land Use Plan.

Fig. 2-1. *If you're planning on a site within city limits, you'll want to visit city hall.*

Fig. 2-2. *Most county real estate development begins at the planning and zoning department.*

THE COMPREHENSIVE LAND USE PLAN

Basically, the Comprehensive Land Use Plan means that the city and county officials have gotten together and decided what type of development will go where. The Comprehensive Land Use Plan is a blueprint to the development of the city and county.

Some states have a State Comprehensive Land Use Plan. The State Comprehensive Land Use Plan, for the most part, acts as a register for the county Comprehensive Land Use Plan. The local Comprehensive Land Use Plan, however, is the plan you need to work with.

To get the overall picture of the entire area, review the Comprehensive Land Use Plan at the county courthouse. Then, if your project is located within the city limits, go to the city hall. Check all of the departments to get the specifics on a site that is within the boundaries of the city limits.

Keep in mind that the Comprehensive Land Use Plan (let's call it the Land Use Plan) is *not* zoning. In fact, you can find various types of zoning in the Land Use Plan that do not correspond.

Two of the main reasons for this are:

1. There are some properties in use allowable under old zoning ordinances. As long as the use of the property stays the same, the zoning board considers this property's zoning to be "grandfathered-in." Grandfathered-in zoning works fine until you want to change the use of the property, or if the building you are using should burn down. In most cities and counties, you could not remodel to any great extent or rebuild if you had a fire. In case of a fire or change of use and/or operation, your new building would have to comply with the new zoning rules and regulations and the Land Use Plan. In addition, these new zoning laws affect setbacks, landscaping, and parking, and you might have a piece of property that is no longer large enough to put an office building or restaurant on. Sound unfair? It happens!

2. There are properties that have old zoning on them, or that fall into an agricultural zoning area, such as farm or green belt areas. For instance, single-family zoning could have future Land Use Planning for multi-family; or agricultural-zoned property could have future Land Use Planning for commercial or multi-family.

   If you have property that was zoned agriculture, but would get commercial zoning, according to the current Land Use Plan, you would not want to rezone until you were ready to use, sell, or develop the property because the real estate taxes on commercial zoning are considerably higher than agricultural zoning.

   Sometimes the city or county will blanket zone or rezone a strip or section of property. For example, a strip of highway property which had previously been agriculture, could be rezoned to commercial, according to the Land Use Plan. This can be triggered by one property owner filing for a zoning change or the city or county could decide that the timing was right and they could use the extra tax dollars.

   In most instances, the old zoning stays in place until the property is sold and the new owner rezones and develops it; or the current owner rezones to command a higher sale price.

The Comprehensive Land Use Plan is usually updated every five to 15 years. The planning and zoning department can tell you how the city or county is growing and changing, and how the Comprehensive Land Use Plan will, more than likely, be affected.

Remember, properties can be rezoned for various uses as long as the new zoning complies with the Land Use Plan. It is possible to change the Land Use Plan, but it can take a lot of time and is not always successful.

It is possible to buy a large piece of property, 500 acres for example, and develop it under a Planned Unit Development (PUD) or Commercial Planned Unit Development (CPUD). This type of zoning and development enables you to go to the city or county with a master plan of how the 500 acres will be developed. This master plan would also outline the infrastructure.

Infrastructure is the utility, road, drainage, and landscaping system and design for the entire property. In other words, the developer must follow the design codes of the city and/or county. The city and/or county must approve the infrastructure design. In most cases, a Directive of Regional Impact study is required for larger pieces of property. The Directive of Regional Impact, also called the DRI, is a study of how your project will affect the environment.

The Land Use Plan is unique to every city and county and should be marked carefully on your work map and, if possible, you should try to purchase a copy of the Comprehensive Land Use Planning Map from the city and/or county.

THE ZONING DEPARTMENTS

Okay, back to work. We are in the planning and zoning department going over the Comprehensive Land Use Plan. We have two basic questions for the zoning department:

1. **Locate existing zoning for your intended use:** Talk with the zoning department and ask them where the property is that is currently zoned for your purpose, what's going on, etc. Any new projects in the mill? Where and what are they? What stage are they in? Is it going through permitting, under construction, etc.?
2. **Can it be zoned for your use:** Where would the property be located in the Land Use Plan that could be zoned for your purpose, but has not been zoned accordingly as yet. Has there been any rezoning activities recently? If so, what and where? Is this area ready for new development? Why? Any new roads under construction or approved for this area? If so, where?

Remember, there are two different categories: the city and the county. Each has its own set of requirements, zoning laws, etc., and can be quite different. If you do not know or have not decided whether your site will be within the city limits or in the county, check both places for specifications.

Mark these areas on your work map, verify the permitted uses for the zoning, confirm what zoning you will need, and also what zoning you could use with a Variance.

A Variance is just what it implies. It is a use that varies from the specific use outlined in the zoning classification. A Variance simply provides the planning and zoning board a way to take an extra look at your project and make sure the development plan you have outlined and your usage of the property won't hurt other developments in the Land Use Plan. For example: A Variance might be given in a commercially-zoned area for a light industrial operation. A Variance helps to tailor the needs of the individual with the Land Use Plan.

A complete planning and zoning rules and regulations guideline manual should be available to you at city hall and the county courthouse for a small fee. Also, summary sheets (which define different types of zoning for you) are usually available in the zoning department. The requirements for setbacks, landscaping, parking, square footage, as well as any other requirements or restrictions, are included in the guideline manual and on the sheets, including other uses that would be permitted with a Variance.

The summary sheets are just that, a summary. Specific requirements not included on the summary sheets are in the planning and zoning manual. The city or county engineer, or one of the zoning officials, should be helpful to you in verifying all of these requirements.

VARIOUS TYPES OF ZONING

Examples of various types of zoning are shown in Appendix B and are a good example of how zoning regulations and restrictions are set up. Each city and county has its own language for zoning. Some cities, for instance, call the commercial zoning B-1, B-2, and B-3, which is Business 1, Business 2, and Business 3. Business is the commercial use and the numerical sequences indicate square footage, parking codes, etc. Other types of zoning include PO for Professional Office, I-1, I-2, and I-3 for Industrial, R-1, R-2, and R-3 for Residential, and MF for Multi-family.

Commercial Zoning

For an example, we will use a commercial zoning that gives you a choice between Commercial Restricted or Commercial General.

Commercial Restricted or CR, includes such uses as shopping centers, restaurants, or motels, and also defines how far back from the front, rear and sides of the property line you can start your paving for parking or driveways, and how far back you can place your building, and anything else you are going to put on your property. These are called setback or building regulations or requirements. They also tell you how much parking you must have, how much landscaping is required, and the minimum amount of highway frontage you must have. Zoning also regulates the height, size and placement of signs. Anything you are planning to put in, under or on top of the property requires proper zoning, engineering and permitting.

Commercial General or CG zoning, includes new and used car dealerships, flea markets, and salvage yards. As you can see, this type of commercial zoning is much broader than Commercial Restricted zoning. This zoning also includes setbacks, parking, and landscape requirements.

At this point you might be thinking, "Hey, I own this piece of property, but yet I have to get permission from someone else as to how I can develop it. Man, I can't even put my building where I want it!" This is an example of the controlled development we are seeing in cities and counties today and it really *is* to your advantage.

Because, while the zoning and planning board is controlling how you develop your property, they are also controlling your neighbor's. A good example is to picture yourself spending a million dollars or more to buy a piece of property, develop it, and build a beautiful building on it—maybe a nice hotel, a beautiful shopping center or a restaurant, and someone pops into town (this is an example with no zoning, remember?) and goes right next door to you, or even across the street, and puts in a dog kennel with 100 barking dogs and no facilities for elimination or control. Or a body shop or welding shop—very noisy with motors pounding and grinding and dust flying everywhere; or some kind of plant with huge trucks running in and out all day, dropping gook everywhere.

The Land Use Plan, and the planning and zoning boards are there for your protection, and in most cases, do a very good job of protecting your property from devaluation, as well as controlling the growth in a city or county. Poor zoning and planning can cause real chaos in a town and can actually lower the value of your property as illustrated in the previous example.

Industrial Zoning

Another type of zoning is Industrial. Industrial Zoning is usually classified as light, medium or heavy industrial, which can also include manufacturing. In many cities and counties, industrial is also broken down into manufacturing categories such as M-1, M-2, or M-3 according to the type of product. For example, a cabinet manufacturing shop would be light industrial, but a steel mill or automobile manufacturing plant would be heavy industrial or manufacturing.

Most industrial areas provide such advantages as railways and ramps, interstate accessibility and, very often, waterways. An industrial area is *designed* to handle production plants with huge trucks running in and out all day, dropping gook everywhere. Not surprising, it would be safe to say that it is also relatively free of complaining neighbors.

It is important to remember that each city and county has its own set of zoning classifications and specifications that can vary, so be sure to verify the specific type of zoning your project needs in the area you are working.

THE FOUR ZONING WHAT'S

The four zoning what's to inquire of include:

1. WHAT'S IN
2. WHAT'S UNDER CONSTRUCTION
3. WHAT'S BEEN ANNOUNCED TO START
4. WHAT'S IN THE THROES OF GETTING APPROVALS

By now your work map might be getting pretty "worked over" looking. In this case, mark the sections of the city or county as Section A, Section B, etc. as shown in Fig. 2-3, and then make notes on your legal pad as to the four What's in each section.

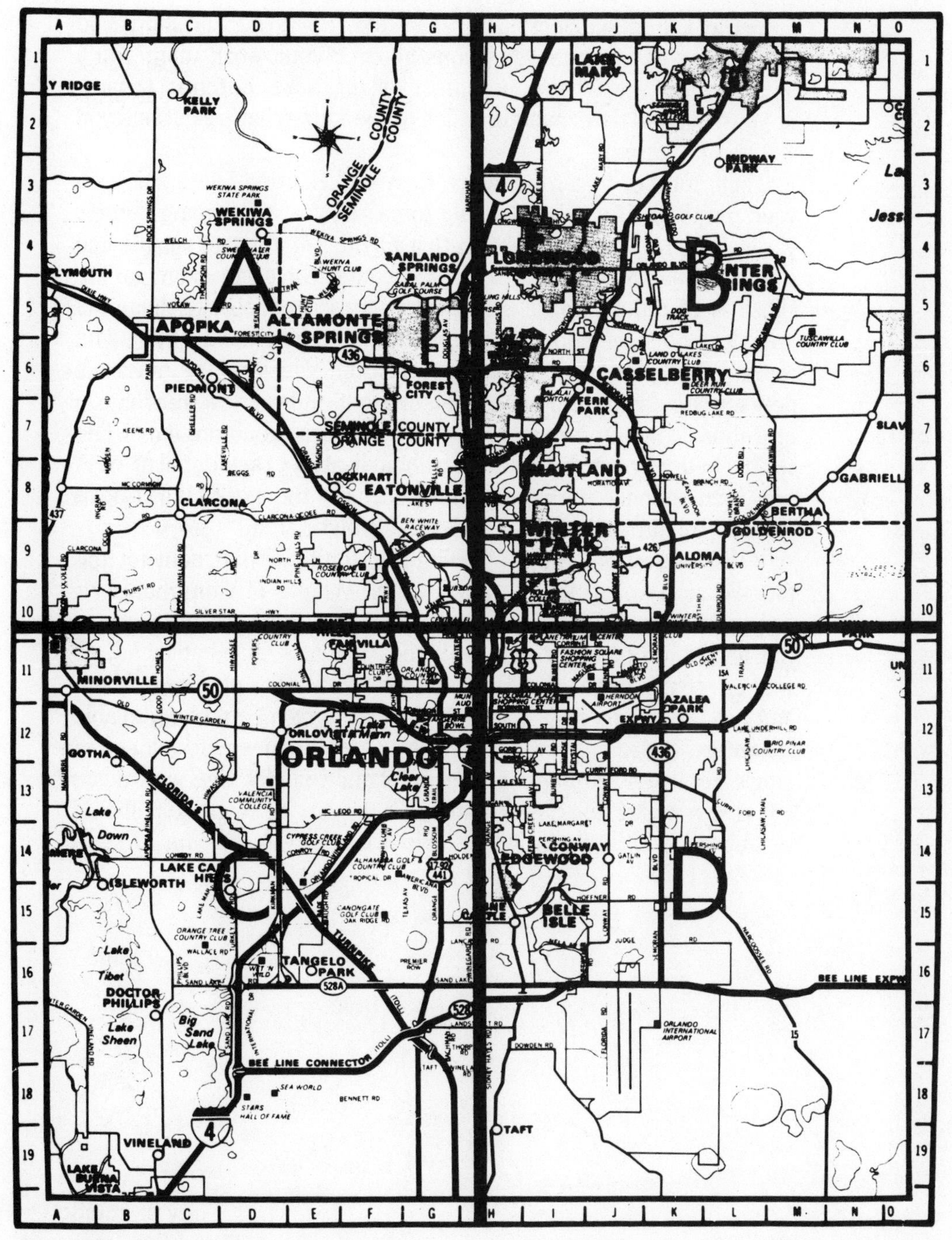

Fig. 2-3. Section off your work map so you can determine the four "What's" in each development section.

Did I mention Sections? I sure did. By now, you should be able to see development Sections all over your work map. You'll see all four of the What's. Your user (client, boss, potential tenant, or yourself) will determine in which of the four What's Sections you will be interested.

Of course, it can be more than one Section, depending on your purpose. If you are looking for a site for a shopping center, you will want to be in a Section that has the amount of people and traffic count you need to attract tenants. This could be in the What's-in Section if there is still a need for your center because the area does not have enough centers to meet the demand. It could be in the What's Under Construction Section, which might indicate a new growth and development area brought on by a new subdivision or subdivisions or, better put, a new "bedroom community. The term "bedroom community" usually indicates a residential area. A new road opening up that provides access to the work place from a rural area, can spur a new residential growth area.

When you asked the planning and zoning department for the four What's, you got it all, not just the What's of Commercial or Industrial Development. Consequently, you should know where the bedroom communities are, if they are new or established, as well as what's been announced and what's in for approvals. Ahah! You now know where property is located, which could be available for future bedroom development in case your user is looking for single-family development property or multi-family. See where you are at this point? You have a handle on the city or town. You should be able to feel the town, where it is and where it is going.

3

Departments

There are many other departments at city hall and the courthouse that can be helpful to you. It is important for you to get all the information you can. The zoning and planning department is where you want to start, but you can't stop there. Use the checklist provided in Fig. 3-5 at the end of the chapter to ensure that you have visited all of the departments and county/city offices necessary.

Let me warn you about a problem with some of the courthouses in areas that are growing exceptionally fast, especially the Sunbelt area. You are going to find that the city hall or the courthouses in some of these areas have literally outgrown themselves. There is so much growth and development going on that the different departments, such as zoning, building, and recording are strung out all over downtown. The building department might be up the street, the utility department two blocks over, and zoning somewhere else. This is a part of the fun of it, and, after all, if the town was not bustling with development, you probably wouldn't even be there.

Keep in mind that the salaries of the people you will be working with at city hall and the courthouse are paid by our tax dollars, and these people are there to help. So don't feel out of line asking them to help you. In any town I've ever been, the clerks, engineers, and officials at city hall and the courthouse have been extremely helpful and very pleasant to work with. A warm handshake of thanks or even a bouquet of flowers is definitely in order.

THE UTILITY DEPARTMENT

What is a utility? These are the water, sewer, gas, electric, and even telephone services you will need on your site. Utilities are critical. At the utility department (city or county), you can check such items as what utilities are available. You'll want to know where the utility lines are, and if these lines are not to your property, who will have to pay to bring them there. Once you get the water, sewer, and gas lines to your site, make sure there is adequate capacity for your project. Also, if the utility lines are already there, *don't* forget to make sure the lines are the right size and large enough for your project. They could have been put in before the present codes were enacted and are no longer adequate.

Some areas in the cities and counties in the growth areas are growing so fast that new water treatment plants and sewer treatment plants have to be built or rebuilt to handle current demands. This could cause a delay in available utilities. Sometimes there is a utility moratorium, which means that there is no sanitary sewer available for your project, or anyone else's for that matter. Water can also be affected and, in some instances, both sewer and water capacity is not available. A utility moratorium can be indefinite. It usually means there is no sewer (or other utility) and no one knows when there will be.

If there is a utility capacity problem, you need to find out how long it will last. Can you go on a waiting list, or make a deposit to assure capacity is available at a future date and what date would that be? Sometimes there is a water or sewer shortage because a new plant is needed. In some areas, natural gas is not available and is so far away that it is unfeasible to bring it in. In this case, propane can usually be used. See the utility checklist shown in Fig. 3-1.

Lift Station

A Lift Station is that not-so-attractive metal box that is on your site or very close to it that provides, simply put, a boost to the

UTILITY CHECKLIST

	SITE			
Service	Size	Location	Capacity	Service Company
City Water				
County Water				
Well				
City Electric				
County Electric				
City Sewer				
County Sewer				
Independent Service				
Septic Tank				
Treatment Plant				
Lift Station				
City Gas				
County Gas				
Storm Sewer				
Telephone				
Other				
Other				

Fig. 3-1. *Use this utility checklist to ensure that you don't forget anything.*

sanitary sewer system as needed. The gravitation of the system and the elevation of the land in the city or county area dictate the necessity and size of a Lift Station.

You will need to find out if a Lift Station is needed. If one is there, check the adequacy, who maintains it and how. Will it become inadequate as a result of your project? Who will pay the cost to upgrade and what will that cost be?

Impact Fees

There is a fee you will be required to pay for your initial hookup to the city or county utility systems. This is called an impact fee. In some areas this impact fee can be very reasonable, but in other areas it can nearly choke you. Generally, impact fees are based on your project's usage and is applicable to every use from a single-family home to an industrial plant. The city or county utility engineer can help you estimate this cost.

Other Impact Fees

If your project is large, or in a new growth area, it is possible that it will trigger the need for more police and fire protection, more schools and more, or improved, roads. This could mean you need to contribute to the costs of these improvements, such as a new fire station or the expansion of police protection. A fairly new impact fee cropping up now is a park or recreational area impact fee. Use the checklist for impact fees shown in Fig. 3-2 as a guide, but ask an official if you have missed anything.

THE ENGINEERING DEPARTMENT

The city or county engineering department can provide information such as what type of drainage you will need, water retention, flood plain areas and any information which would be pertinent to the development of your site.

These things can be very important and affect the size of the property you buy. If you need a large retention area, some extra

CHECKLIST FOR OTHER IMPACT FEES

	Adequate	Inadequate	City	County	State
Police Protection	______	______	______	______	______
Fire Protection	______	______	______	______	______
School System	______	______	______	______	______
Road System	______	______	______	______	______
Parks or Recreational Areas	______	______	______	______	______
Other	______	______	______	______	______

Fig. 3-2. *Checklist for impact fees.*

drainage, or your site is in a flood plain, then you might need a larger piece of property than just what you need to stick a building and some parking on.

Retention Area

An undeveloped piece of land, whether it be one acre or 100, has a certain amount of water that is absorbed into the ground when it rains or during a storm. Once this property is developed and a building has been built and asphalt parking put in, it no longer will hold and absorb this water.

In order to prevent this water from draining to neighboring properties, or causing flooding-type problems after it is developed, the developer could be required to put in a retention area which will hold and absorb this water. See Fig. 3-3. The engineering department has the formula as to how deep and wide this retention area should be and how much water it needs to hold and for how long.

Some areas have a different system set up such as central drain canals, in which case property development can be designed to use.

Fig. 3-3. *Retention areas hold water that is not absorbed by the property and controls water runoff and flooding.*

Detention Area

The detention area is designed, like the retention area, to absorb rain water on developed property. Unlike the retention area, which is designed to be wet and usually must be fenced, the detention area or areas, are sloped and graded to absorb the rain water and not hold it. For instance, the detention areas in an apartment development could be graded and landscaped areas running through the project development. This could be designed to look like the natural roll of the property.

Flood Plain

In some areas, the elevation or sea level is such that there has been a flood in some areas of the city or county, or a flood is predicted. This area is called a flood plain. See Fig. 3-4 for a sample of a flood plain map. You need to keep in mind that property located in a flood plain needs to be developed very differently. Less of the property is developable due to the drainage and absorption level. The engineering department has the information you need regarding the flood plain.

Ordinance

The city or county engineering department should also be able to make you aware of any ordinances that will affect the development of your site. An ordinance is a specific rule or law, and in development, usually further defines just how the planning committee wants particular issues handled. For example, some towns want to save as many trees as they can in order to protect the atmosphere and landscaping that made it attractive to begin with. Sign ordinances are also a big issue today. Because of rapid development and everyone wanting a bigger and more visible sign, city and county planners can see that without a sign ordinance, they could wind up with a city of unattractive and even offensive signs.

State and Regional Agencies

There are state and regional water and environmental agencies you could need approvals from. While you are in the engineer-

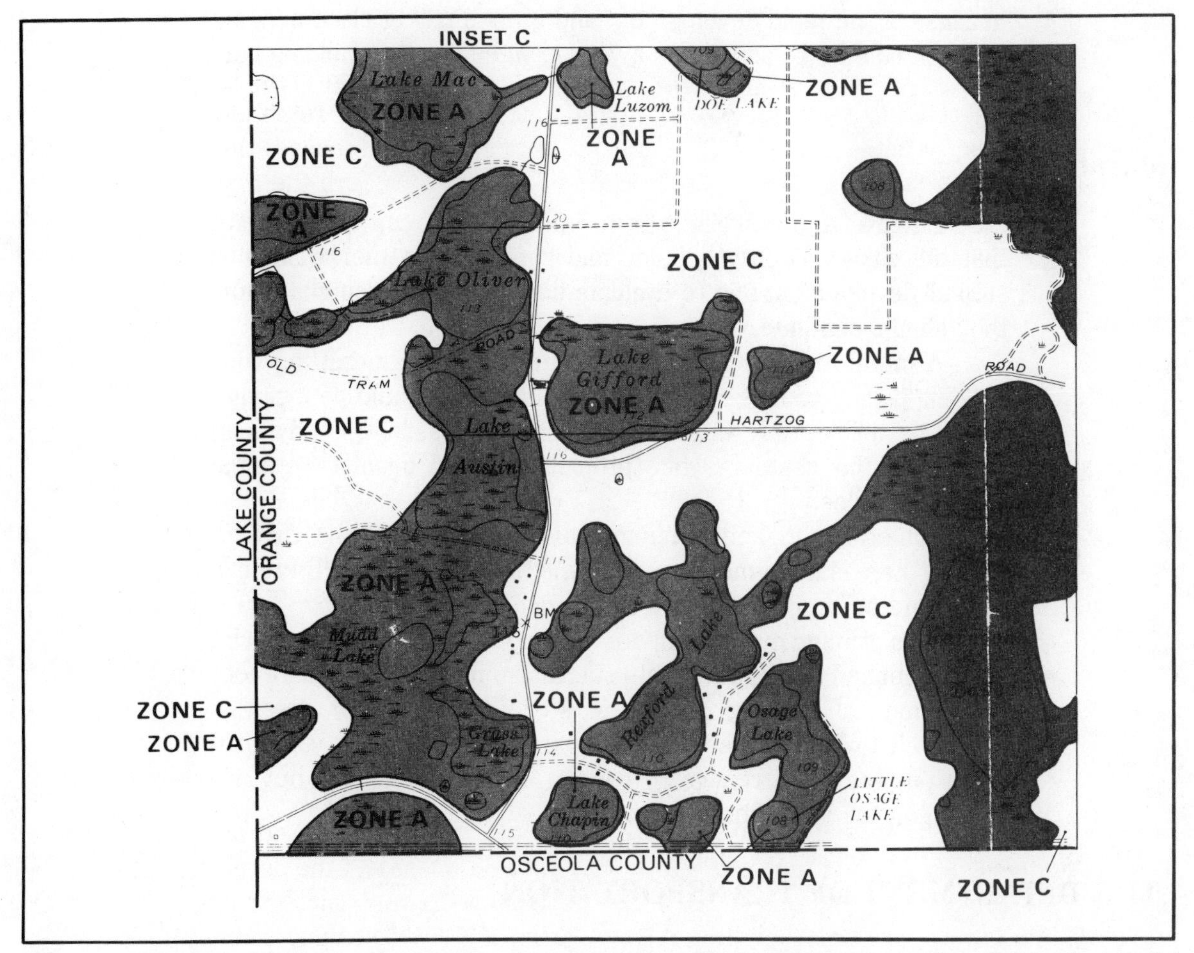

Fig. 3-4. This flood plain map shows areas that are at risk of flooding, which determines how a site is built on.

ing department, ask the engineer you are working with what state or regional offices would be involved with the site and type of development you are planning.

The bigger your project the more permits you will need.

THE BUILDING DEPARTMENT

The city or county building department can assist you with information you will need for your building permit. You cannot build, clear, pave, or remodel without a building permit. Again, because of the rapid development and growth taking place in some areas, it could take as long as a year *or more* to get a building permit.

Moratoriums

Be careful of building moratoriums. Things can grow so fast that the city or county building and zoning departments have to stop all development and re-evaluate how things are going and how they should continue.

A moratorium or an extended period of time in getting your building permit could be very costly to you and should be included in the feasibility study as a holding cost. The feasibility study is an outline of the costs in your project matched against potential income or sales.

Something to keep in mind about moratoriums is that although there is not a moratorium at the time you inquire, there can be a moratorium planned for a future date.

This planned moratorium for building or utility capacity can be predetermined by different facts such as so many more square feet of commercial building allowed or so many gallons of water or sewer left to be allocated. Be very careful about this as you could close on your property, go to get permits, and they ran out two days ago. Ask lots of questions.

THE DEPARTMENT OF TRANSPORTATION

The department of transportation is a state office and is not located at city hall or the courthouse, but is in a building of its own. Check the yellow pages or ask while you are at the courthouse where it is located. Your research would not be complete without a visit to the department of transportation (DOT).

I worked as a broker with a developer who had to walk away from a beautiful piece of property on which he had planned a residential development. After checking with the department of transportation, he found that the existing road in front of the property would have to be repaved and brought up to DOT standards to handle the increase in traffic the new development would generate.

The cost to repair and resurface a mile and a half of badly

worn county road made another site more feasible for the developer. Not only did this add to the cost of the project, but also would have cost the developer too much time. He needed to have his development ready for the spring market and could not meet his deadline because of the road.

DOT will be able to tell you if there are any new roads planned, any widening or resurfacing of existing roads, ingress and egress information (how to get in and out of your site), if a traffic signal is planned or will be needed as a result of your development, if you will need to put in deceleration and acceleration lanes, and a wealth of other information about how your project relates to the current and future traffic flow. Deceleration and acceleration lanes are those paved strips on the shoulder of the road for a car to pull into when slowing down to turn in, or to pull out into, to pick up speed before entering the flow of traffic.

Road impact fees based on how much your project will increase traffic on the existing road could be applicable in some areas.

Ask lots of questions at the department of transportation and ask the traffic engineer if there is anything you have missed. He can help you make your project a success.

THE TAX ASSESSOR'S OFFICE

There is another office—the tax assessor's office or public records department. In this department, you can find the name of the owner of the property you like or are interested in. You can obtain the property owner's name and address, the legal description of the property, what he paid for it and when, the assessed value and the amount of the real estate tax applicable to this piece of property. We will come back to this department after we take a look at the sites themselves.

OTHER DEPARTMENTS

Don't limit yourself to only the departments and informative sources I have mentioned here. These are the departments and offices you cannot overlook.

CITY HALL/COUNTY COURTHOUSE CHECKLIST

Site:______________________________
City:________________ County:________________
Size:________________ Dimensions:________________
City Limits:________________ County:________________
Comprehensive Land Use Zoning:________________
Present Zoning:________________ PUD:________________
Variance:________________
Permitted Uses:________________
Departments:
Building:________________
Square Feet Allowable:________________
Moratorium: Present________________ Planned________________
Time allowance for building permit________________
Special City/County requirements________________
Engineering:
Elevation of site:________________
Flood Plain:________________
Retention/Detention Requirement:________________
City/County Drainage problems:________________
____
Ordinances:________________
Tax Assessor:________________
Property Owner:________________
Address:________________
Phone Number:________________
Legal description:________________
Assessed Value:________________
Present owner's purchase price:________________
Date of Purchase:________________

Fig. 3-5. *Use this city hall/county courthouse checklist to ensure that you've visited all of the necessary departments.*

Utility:

Describe location, size of lines, what capacity is needed and available

Sewer:________________________________

Water:________________________________

Gas:________________________________

Telephone:________________________________

City/County problems or special requirements________________________

____

____

Moratoriums:

Building:________________________________

Utility:________________________________

Other:________________________________

Department of Transportation:

New roads:________________________________

Resurfacing or widening planned:________________________

Ingress and/or egress availability:________________________

Fig. 3-5. *(Page 2 of 2)*

If you are not sure which department has the information you need, stop into any of the offices at city hall or the county court-house and tell them what you need and get directions to the department that has the information. Keep your eyes and ears open to any other department or agency that can be of help to you. Ask lots of questions, and use the checklist in Fig. 3-5 as a guide.

4

Feel, See, and Touch

I want you to Feel, See, and Touch the town or area you want to develop your site in. Don't leave the courthouse until you Feel the city or area you are working. Make sure you can picture that ol' Land Use Map in your mind and how it works and why it's there. You should feel like you have a crystal ball. And, if you've done your work right so far, that's exactly what you have.

SEE

It's time to See. We can now drive around and look at the town and see the Land Use Plan in action. Plan your travels through the sections and areas of your city or county in a very organized fashion. Closely follow what you are doing on your work map. Be sure to See everything you have marked on your map and make notes as you need to.

In fact, you should travel these areas until you can close your eyes and picture every one of them. Picture a Land Use Overlay on each of your mental pictures. Don't rush, be organized and use all the time you need. These steps are so important.

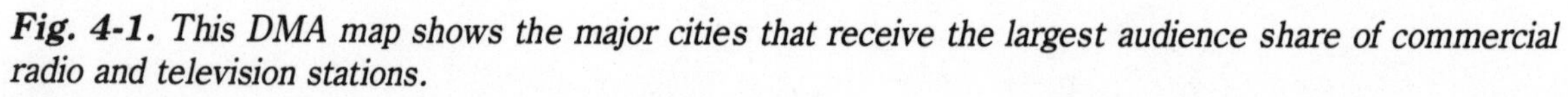

Fig. 4-1. *This DMA map shows the major cities that receive the largest audience share of commercial radio and television stations.*

Nielsen Media Research

While you are looking in your sections, keep an eye out for sites. You might want to drive through each section or area and then, after you've seen all of them, go back and check for sites. If you want to penetrate the market with a particular user, you will want sites from each section, and if you want only one site, you need to pick the section of development area that best suits your purpose.

We will get back to Touch when we contact the owners of the properties we find. We will Feel the city, See it and Touch it by contacting the property owners.

Let's talk for a minute about market areas or television coverage areas. Some clients or companies call these designated market areas (DMA) or area of dominate influence (ADI) see Fig. 4-1. The DMA or ADI is generally a group of counties where the commercial radio and television stations in the metropolitan and/or central area, achieve the largest audience share. This is non-overlapping geography for planning, buying, and evaluating television audiences across various markets. In other words, this shows how many cities and counties pick up the radio and television stations. If you are going to penetrate an area, the area in which to concentrate is the DMA or ADI. This allows you to get the most from your marketing efforts and dollars.

If you have two tire stores that are in different cities or counties but in the same DMA, it would save considerable advertising dollars because one ad would handle both towns. But, on the other hand, if your tire stores were in cities or counties located in two different DMAs, it would be necessary to buy two ads, one in each DMA. Your advertising efforts and dollars could double.

Market penetration is done by establishing the market you want and putting enough of your tire stores, strip centers, child care centers, or restaurants in that area to establish an identity, as well as supply the demand or need for your product throughout the entire market.

There are two approaches you can take: Your area could include one city or county, or a DMA. Each city or county, as we have discussed, has sections in it. Therefore, even if you are working in a DMA, you want to take each town in the DMA, or each major city or county large enough to meet your development requirements, and research it. Every city or county has a courthouse and a city hall with all the departments needed to put your project together.

GETTING "IN TOUCH"

The tax assessor's office has the name and address of every property owner in the city and county and the legal description of the property. Back to the courthouse. Although this is a county office, it also includes property owners located within the city limits of the various cities in the county.

Bring along your work map, and identify on the map where the site or sites are that you want to check. You might want to do this a section at a time, depending on what you are looking for and what is available.

The tax assessor's office or the tax appraiser's department also has the purchase date of the property, the purchase price, the assessed value, the real estate taxes, and any improvements made (such as buildings, driveways, etc.). See Fig. 4-2 for a small tax assessor's printout.

In some area, this information is stored in computers, which you are welcome to use. Again, the clerks are usually available and willing to help you find the information you need (including how to work the computer).

The site is identified by the legal description, but if you don't have that, check the location on the map. The tax assessor's department, as well as most of the other departments, will have a large map that will enable you to determine the location and legal description of the property. Any identifying cross streets or improvements such as a convenience store or bowling alley will help determine the exact location of the site you are checking.

If the site in is a Planned Unit Development (PUD) it might have a subdivision number, especially if it is within the city limits. Subdivisions and their numbers are usually shown on the plat books and maps. Ask for help in finding where to look for this information.

In many counties, the tax assessor's office has a map showing the sale price of property sold within the last two to five years.

```
                         TAX ASSESSOR'S PRINTOUT


 EXEMPTIONS
SALE03/80 PRICE   173000ORB 473 251AFD        V WD O



  R17 25 29 0000 0055 0000
LOC  2401 W  IRLO BRONSON MEM HCTY03TDS 0303   SF WF        KEY  372698
----MAILING ADDRESS------   PCA2711      PCS00    YR78    PARENT       0
KAY WHITEHOUSE              MAP000      AREACM16  JV         MTG0000
P. O. BOX 38662             SP1          SP2            SP3
                            UT1          UT2          SQ FT  3850
ORLANDO         FL 32811    AYB1977      EYB1977 OBS     CONST
                    0000    LAND     191250   IMP       47408 OTHER        3063
----LEGAL DESCRIPTION----   TRUE MKT    241721  REA       CLASSIFIED
FROM SE COR OF SEC, RUN N   ASD LND   191250 ASD IMP      47408 ASD OTHER    3063
80 FT & W 1321.29 FT FOR   DESCRIPTION   TAX YR   CURRENT     EXEMPT          TAXABLE
POB, N 210 FT, W 125 FT,    TAXABLE VL    202176     241721                    241721
S 210 FT, E 125 FT TO POB   SPEC DIST     202176     241721                    241721



708256      SPLIT060684
 EXEMPTIONS
SALE10/85PRICE    450000ORB 7861595AFD        I WD U
```

Fig. 4-2. *A sample tax assessor's printout including the purchase date of the property, the purchase price, the assessed value, the real estate taxes, and any improvements made.*

This is usually just the commercial areas. This map is great in comparing property prices, but make sure it is up-to-date. The tax assessor's office should have information on all sales in the city and county.

When you get the name, address, and telephone number of the property owners, you can begin to contact them regarding their property.

> **NOTE:** If you are trying to find a site for a smaller user like a bank or restaurant, some of the sites you will find interesting could be out-parcels to shopping centers or a development. Because the property could still be under option, and the sale has not gone through yet, the

tax assessor's office will not have the information you need to get in touch with the new owner or the person who has it under option. The developer or marketing agent's name and telephone number are usually on a sign at the site. If not, and the project is under construction, the building and/or engineering departments of the city or county will have the name of the developer. In some cases, the previous owner might provide you with information to get in touch with the new owner or developer. Ask around, don't give up; someone is going to know.

If you are nervous about calling the seller cold turkey and talking about their property, practice what you are going to say with your spouse or a friend. Have them play the role of the property owner or developer. Practice calling them. Tell them you are interested in purchasing the property they own in Wherever, USA.

It's a good idea to practice what you are going to say to an owner or developer with a friend before you call.

You'll be surprised how this exercise smooths out your conversation and sharpens your communication skills.

In some cases, it will take more than one property owner to put your project together. Always have an attorney assist you with contracts and offers.

SECTION 2

5

The Time Clock

After you have identified your area or areas, the next step is finding the best site in each area. I am only going to talk in terms of A and B sites and mostly about A sites, A being the best. Being the best is the name of the game today, and mistakes can be very costly.

There are many factors that go into making a site the best. We will go over these factors and how to add and subtract positive and negative features to meet our requirements.

Let's talk about time, the twenty-four hours in each day that you and I have to do the things we need and want to do. If you have enough of this scarce commodity these days, you're either extremely organized or you're doing something wrong. Customer service businesses (and aren't they all) need to plan on all of their present and future customers being *very* time conscious.

I'm not talking time of the day, or even how much time our customers have to make one stop. Today's customer—the wealthy to the low-income; the youngster to the grandparent; the career-minded to the retired; families to singles; and just about anyone you can think of, is not thinking of making one stop or having one thing to do. We are all suffering from the time crunch. When a customer has a stop to make, he is going to head for the destination point or commercial impact area where he can get the most done.

Keeping this in mind, and picturing that clock ticking away without a care, let's see how a few people like ourselves handle the time crunch.

THE WORKING MOTHER

Lavonne leaves her office building or place of work, goes by the day care center to pick up the children, stops at the drug store to get Valerie's ear medicine, the shoe store to get Johnny's tennis shoes, and the sports store to get Dad some golf balls. Next, she goes to the department store or discount store for socks for everyone and a little camera for the picnic tomorrow (film too, of

Today's working mothers want to get errands done quickly, and more often than not, shop in one conveniently located place.

course). By then, the kids are getting rowdy because they are hungry so they all have dinner. During dinner, Mom says, "Look kids, the new Disney movie we've been wanting to see is playing here at the cinema. Since Dad is fishing tonight, let's go," (unanimously agreed upon by all, of course). Off they all go to the movies, and last but not least, stop by the grocery store on the way home and get groceries. Now, because everything was right there in one place for mom, all her errands have been run and she can take the kids on a picnic tomorrow—Saturday.

THE TEENAGER

"Hey Mom, can I borrow the car? I want to get my hair trimmed, drop my shoes off for repair, get some of that cool psychedelic eye shadow, pick up some vitamins for the kitty, look for some new jeans to wear to Sally's birthday party tonight—oh yeah, I need a birthday present too. You need the car at four o'clock? Gosh, that's two hours from now, it won't take me that long. I'm going over to the shopping center where everything is right there together. Yes, I'm sure I can be back in time. Okay, I'll pick up some milk and O.J. Can I get some hot dogs too?"

Teens, like adults, are busy with work, school, homework, activities, and friends.

This teenager even had time to make a sneak stop by her girlfriend's house to see her new sweater for the party and still got home in plenty of time for her mom to have the car at four o'clock.

MR. BUSY EXECUTIVE

Jim leaves his office complex and drives straight to the gym. After a forty-five minute workout he showers, dresses casually, and takes off for the service area (destination point) where he can get the most done in the least amount of time. There's a party at seven o'clock at the Smith's and his wife Nancy will kill him if they are late. Conveniently dropping off his suit at the dry cleaners, he picks up his daughter's shoes she dropped off Saturday. Then he stops at the drug store for shaving cream, picks up a birthday card for his sister Theda, a new tennis racket, and plane tickets for his trip Monday. He also gets the oil changed in the car in about twelve minutes, gases up, and picks up a video for the kids to watch while he and Nancy are at the party. Last but not least, Jim picks up a gallon of milk and flowers for Nancy.

Today more than ever, a working man's time is limited. More than likely, he runs his own errands and helps out with the household shopping and chores.

THE LUNCHTIME STAMPEDE

"Okay gang," announces Roger, "we've got a car leaving at exactly 11:59. We can hit all of the stores because we are going to the new shopping area where it is all conveniently located. Be late and be left."

Everyone piles into the car and arrives at the shopping center within 7 minutes. Everyone heads off in different directions to tend to their errands.

Sue stops by the 24-hour teller for some cash, picks up her tickets to the Reba McEntire concert, has her eyes examined and orders her glasses so she can pick them up on her way home tonight.

Kevin needs to order wallpaper, pick up the shirts he had tailored, order flowers for his secretary's birthday, and then picks up his car that he dropped off this morning at the window tinting shop.

Party favors, gift items, household items, hair cuts, car payments, sport supplies, manicures, insurance papers, even a loan closing are all completed during this lunchtime stampede.

Many people use their lunch hour to run errands while they grab a bite to eat.

At just the precise moment, everyone meets back at the car and heads for the drive-thru window of the fast food restaurant. Since Jerry is driving, he gets to pick the fast food restaurant the shoppers hit today. A quick stop at the drive-thru window brings everyone back to work with errands run and something in their tummies.

This really happens. I have been on a stampede or two myself. It is the only way to get everything done when you work all day, all week.

The lunchtime stampede is a highly-organized, precisely timed operation. Only the serious, time-conscious errand runner should participate. Proceed with caution and make sure your insurance is paid up.

RETIREES, SINGLES, AND CHILDREN

There are so many fun things to do now for retired people that they also are time conscious. And single people, as well as

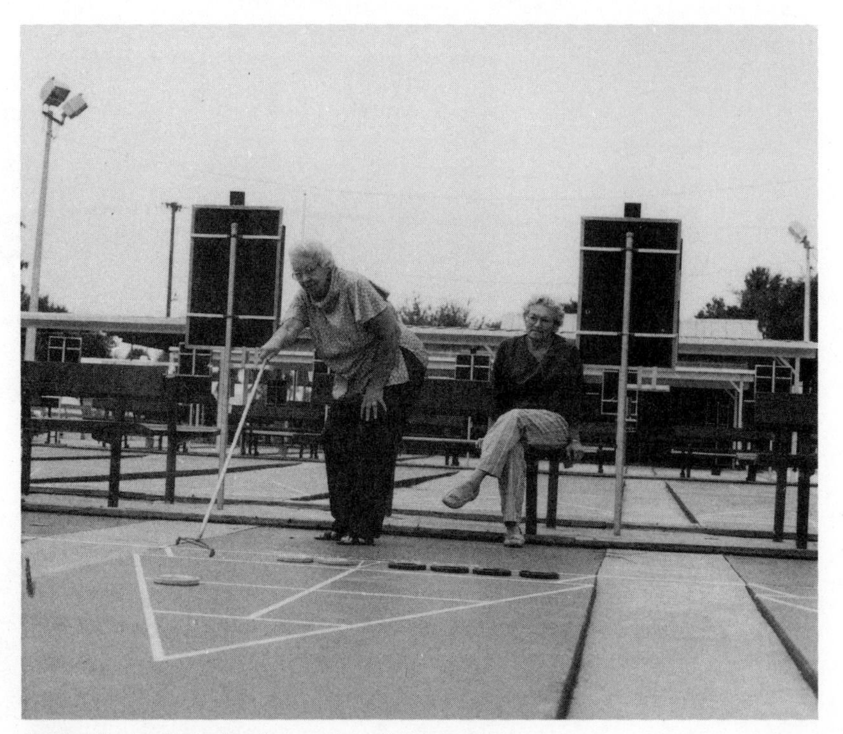

Retired people today live active, healthy lives.

retired people, have things they want to do. A single person has to juggle between work, social activities, domestic chores, college classes, and visits to home. Children also have just so much time to attend school, do chores, complete homework assignments, and of course, play. Why, everybody is busy.

If you want to select a site that is going to fit into the best category, keep time on your mind. Every project we do is a customer service business. Our customer might be a client, tenant, or employee. Save your customers time. For example, your client can get to you easier and faster than any other restaurant in town (of course you will also need decent food); your tenant can provide his customers plenty of parking, easy ingress/egress, and is in his path of everyday travel; and your truck driver has good access to a major road artery so deliveries are on time. The precious commodity of time can be used to your advantage and give you and your site that valuable distinction of being the best.

6
Needs

By now you should know where all the people live, where they work, how they get there, and their level of income—and anything else pertinent to your need.

If you were looking for a property that was large enough for a shopping center, it must have the population, income level, age group, and traffic count for your tenant's needs.

You need to know what population the tenants you are going to target must have to do the volume of sales they want. All tenants have customer income levels they appeal to. For example, you don't want a shopping center with tenants such as Sax and Tiffany's in a low- to middle-income area. If you are planning a large regional mall, however, you could strategically place your center for the low- to high-income population and have a large enough tenant base for everyone.

What I am saying is, in order to find a site for your client or boss, you've got to know their needs and requirements. Council with your client and determine what his needs are. If you are employed by a company, determine for yourself what your company's needs are.

Needs are requirements such as the size of the property, the dimensions, traffic count, population, and topography.

Every company and every user has their own set of needs and requirements. If they are completely different from the ones we discuss, make your own list and define exactly what they are.

SIZE

What size does this piece of property or site need to be in order for your client or boss to develop it, build his product on it, and satisfy all requirements such as building codes, set backs, surface drainage, and parking.

As mentioned in Chapter 14, you need an architect or an engineer, and sometimes both, to determine this for you. Your architect can calculate the size of your building or improvements, how much parking, landscaping, etc. you will need, and should be able to give you a good estimate of how much property you will need.

Your personal preferences could include extra width for visibility, or more trees for atmosphere. If you want more trees, this could very likely mean you will need more property, (trees take up space you know). I was trying to put a restaurant on a site in Osceola County in 1987, and one beautiful oak tree took up twelve parking spaces.

Trying to squeeze your project onto a site that is too small can be a disaster. See Chapter 13, Fifty Flags to Site Selection.

Of course, it goes without saying that, if the site is way too large and cannot be divided, you or your client could be paying too much money unnecessarily.

DIMENSIONS

Along with size, some users have a particular set of dimensions they need to make their concept work. For example, if your user is a restaurant, and they want to place their building in the middle of the site so that parking can be located on all four sides of the building, you are going to need a site with more width or front footage. If you need a drive-thru window (check the zoning—I've run into some areas with zoning that prohibits drive-thru windows) your width needs to accommodate the drive around area.

If your building is long, the depth of the site could also be a critical requirement. For instance, mini storage projects prefer a deep, narrow lot. This cuts down on the cost of paying for extra

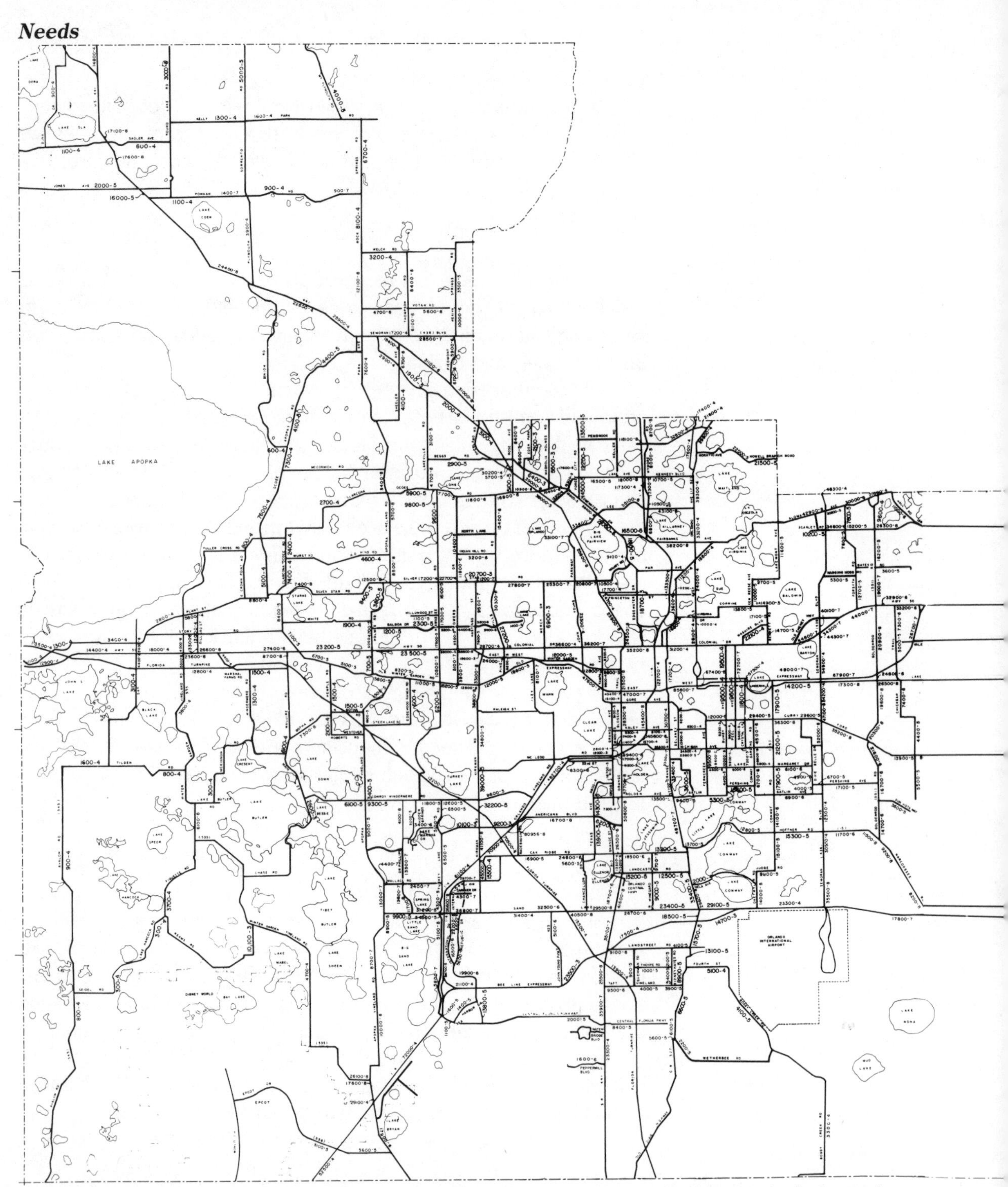

Fig. 6-1. *A traffic count map like this one can help you determine if your site is strategically located.*

ORANGE COUNTY 24-HOUR TRAFFIC COUNT MAP

EFFECTIVE DATE: MAY 10, 1988

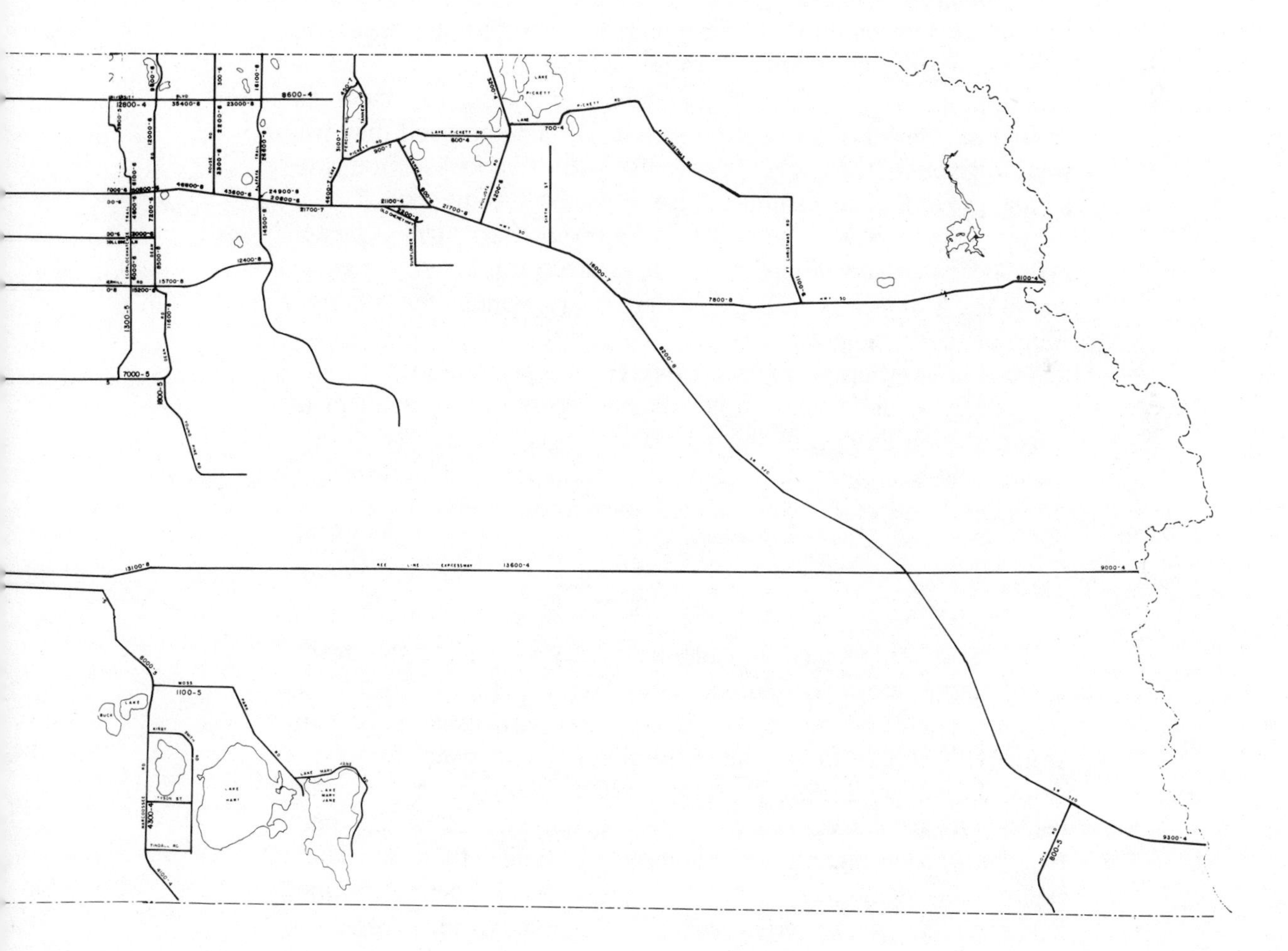

frontage and enables them to build their storage units in a long, narrow building.

Shopping centers usually require excessive frontage exposure, and some users insist on corners. President Lincoln must have had real estate development on his mind when he stated how difficult it was to please all of the people all of the time.

TRAFFIC COUNT

According to statistics, it takes a certain determinable number of vehicles to pass your door each day to generate a particular number of sales. Of course, the type of user makes a difference. For example, a bookstore would have a different requirement than a fast - food restaurant. Traffic count is important because you want to be on your customer's route.

It is easy to say, "We will have plenty of reasons to draw them around the corner and down the street. We'll be so good they will want to go out of their way to get to us." Don't be foolish. You have a great project, your tenants are great, you have some real winners for your out-parcels, or you have an excellent cherry pie—why not take all this great stuff and get a great site for it? I don't want to sound like I'm lecturing, but don't create a mountain for yourself. Eliminate every negative you can. The competition is fierce. Don't underestimate them and **remember the time crunch**.

Traffic count maps can be obtained from the department of transportation as well as independent companies that do market studies and demographic studies. An example of a department of transportation traffic count map is shown in Fig. 6-1.

POPULATION

The demographics of a city or county as well as the designated market areas (DMA), can be obtained from the chamber of commerce, or in some of the material you purchased at the map store. This is okay to get you started, but you will want to order a detailed demographic profile from one of the demographic companies. If you have not worked with one of these companies, ask them to send you an example of the type of studies they do. Their report—the demographic profile—should show the population, household income, age, racial mix, ratio of homeowner to tenant, a breakdown of the types of employment, (executives, clerical, laborer), and other information that is very important for you to know and will add credibility to your presentation. Refer to Appendix B to see a sample demographic study.

TOPOGRAPHY

Topography is the details of the physical features of property. A topography map shows such details as lakes or ponds, location of trees, including their size and what kind they are, and the elevation of the site.

A specific user might want something particular in the land. A developer for a single- or multi-family project might want trees or a small lake; a Marina needs a large lake or river (one on which he could get a permit of course).

A unique example of utilizing topography is a little strip office and retail center on State Road 520 in Cocoa Beach where I worked many years ago. Among other tenants such as the insurance office where I worked, a real estate office and an office supply store, was a small retail boat store. This little office center just happened to back up to the Banana River. When you went to the boat store, you could actually try out the boat you wanted, buy it, and drive it away.

I am not sure you could do that today because zoning laws have changed, but there are some areas where boat sales are set up the same way, on a much larger scale. For instance, large marinas like to have retail boat sales as part of their operation.

Another example is a hunting lodge. You would need a lake, some trees and bushes, and of course, some game stock would also be helpful.

STRATEGIC LOCATION

The Going Home Side

Sound strange? You are more familiar with this need than you might think. Have you ever been going home from work and wanted (or needed) to pick up a prescription, a pizza, etc.? You are in your car—in traffic—and the drug store or pizzeria is on your left. This means you will have to change lanes, turn left (sometimes even go down to a traffic signal or median cut and double back) and cross traffic to get to what you want. Let's paint another picture. The drug store or pizzeria is on the right side and all you have to do is slip into the deceleration lane, turn in, and slide right into a parking spot. If it is a destination area, the drug store and pizzeria will probably *both* be there.

Some users feel that it is critical for them to be on that "going home side" of the road. This is the side most of the traffic will be on because that is the best route to take home from work. (Sometimes the *only* route home).

The types of users who would want to be on the "going home side" might be restaurants (especially fast food restaurants), or banks. If there is a strong destination area between work and home, the traffic (customers) might be just as happy to go there if they have more than one stop to make. Remember The Working Mother we discussed in Chapter 5.

The Morning Side

The same idea applies to the morning. If you want to drop the kids off at a day care center or have breakfast, you will want these places to be on the right side of the road in the morning. Determine which side of the road your client will need. A strategic location will give you that extra edge you need on your competition.

Tomorrow, when you are driving home from work, check this out. See what benefit a bank or restaurant on *your* going home side would have. What would you need on your morning side?

A GENERATOR

It might be important to your operation or project to be close to what is called a generator. This is a business, shopping center, office park, etc. that generates people and attracts business for you. A shopping center is a customer generator for a bank or restaurant because it draws people to the area or destination point.

These points bring us back to the time crunch again. The fact that we are all so short of time makes it possible for one business or store to generate business for another. An excellent example of this is the grouping of hospitals, x-ray laboratories, and doctor's offices. They all feed off each other and the patients, clients and even the suppliers, benefit from the convenience.

OTHER NEEDS

Other needs include zoning, types of utilities needed, roads, railway or waterway availability, maybe an airstrip, good ingress and egress, and visibility.

Have you noticed how much we are talking here about *need*? Need, need, need. You, your boss or client have specific needs. There are specific things from this site that you need. You have to satisfy as many of these needs as you can—preferably all of them.

When you find yourself saying, "Well, maybe I don't need this, and maybe I can get by without that," you better stop and determine what these "give-ups" might cost you. Will it cost you sales because people can't get to you? Oh yes, they can see you, but maybe they can't get to you. You can't get tenants because you are around the corner from the main "drag"; your drivers can't get in and out and you are losing orders because you keep running late. You can't take these chances.

When you skimp on size, give up a median cut (that break in the grass or concrete strip in the middle of a four-lane road) or turning lane, lose visibility, develop next to a sewer plant, give up density (number of units allowed by zoning per acre), decide not to buy that traffic signal, forget the time saving rail access—it is the beginning of the end my friend. Your boss is going to either think you are an idiot and fire you; or take your advice and fire you after the company loses money on a bad site. Your client will realize you are not the professional he thought you were or if you are investing for yourself check with your mother and see if you can get your old room back.

There's an old real estate cliché that says the three most important things about real estate is location, location, location. This can be true, if you add to it better, better, better. The way to make your site better is by satisfying as many of your needs as possible. As I said before, preferably all of them. And, always remember the time crunch.

If you have your area picked out and the only site you can find that works is not the greatest you have seen, you need to analyze this site very carefully. If the site is for your client or boss, bring out the negatives with the positives and let him make the decision. If your client or boss wants your recommendations, give them to him based on how you feel after spending time in the actual area.

Give him what remedies you might see to the negatives, if there are any, and provide the clearest picture you can of all the details. Never try to sweep the negatives under the carpet. Bring them out in the open and address them. Sometimes you will find a diamond in the rough.

The best site has exactly what you need. Be creative however, if some of your needs are not there maybe they are available. Perhaps a turning lane could be negotiated with the department of transportation. Although you have to pay for it, it could make or break your project. Put the cost in your feasibility study and see how it factors out. If you need a lake, you may be able to dig one.

Analyze your needs and prioritize them. You will discover that there are some things that are impossible to do without. For instance, you are going to build an office building on your site, but when you check with the zoning department, it is classified for single-family homes. Your first thought would be—maybe I can have it rezoned.

Back to your Land Use Plan; its says single-family homes or multi-family with low density with no variances for office building. You even go a step further and check with the city to see if you have a chance at all in getting this rezoned for your use; but, they say absolutely not. You have a site with a great location, however, it doesn't fit your needs. There is nowhere you can go with this site—it is over.

CHECKLIST

We have discussed our needs—size, dimensions, traffic count, population, topography, zoning, utilities, roads, and waterways. Now make yourself a checklist specifying your needs. Rate your needs something like this: SIZE: Best, Will Work, Getting Scary, No Way. An example of a checklist form you might use is shown in Fig. 6-2. The checklist is not in any order of priority. You might think of things you want to add to your checklist or you may want to use different categories. Do this for each of your sites, and when any of your needs get to ''No Way'' it kills the site and you start with the next one. If you get two ''Getting Scary's'' evaluate how important these ''Getting Scary's'' are to your project. If they

SITE CHECKLIST

SIZE	Best_____	Will Work_______	Getting Scary_________	No Way_____
DIMENSIONS	Best_____	Will Work_______	Getting Scary_________	No Way_____
TRAFFIC COUNT	Best_____	Will Work_______	Getting Scary_________	No Way_____
POPULATION	Best_____	Will Work_______	Getting Scary_________	No Way_____
DEMOGRAPHICS	Best_____	Will Work_______	Getting Scary_________	No Way_____
INCOME	Best_____	Will Work_______	Getting Scary_________	No Way_____
AGE	Best_____	Will Work_______	Getting Scary_________	No Way_____
ZONING	Best_____	Will Work_______	Getting Scary_________	No Way_____
UTILITIES	Best_____	Will Work_______	Getting Scary_________	No Way_____
DRIVEWAY	Best_____	Will Work_______	Getting Scary_________	No Way_____
PRICE	Best_____	Will Work_______	Getting Scary_________	No Way_____
SOILS	Best_____	Will Work_______	Getting Scary_________	No Way_____
_____________	Best_____	Will Work_______	Getting Scary_________	No Way_____
_____________	Best_____	Will Work_______	Getting Scary_________	No Way_____
_____________	Best_____	Will Work_______	Getting Scary_________	No Way_____
_____________	Best_____	Will Work_______	Getting Scary_________	No Way_____

Fig. 6-2. *Use this site checklist to analyze whether the site you've chosen is the right one.*

are important (and if they are not, they shouldn't even be on your list) kill the site.

Getting the picture? Stay in the "Best" or "Will Work" categories. Do you want to spend your money on a "Getting Scary" site? Do you really want to present a "Getting Scary" site to your boss or client? Go for the A sites—go for the Best. The competition is ferocious. Be Better, Better, Better.

7

The Competition

Oh no, the competition is here! What should I do? Well, first of all, what is competition? It is the "other." The other person, the other restaurant, the other day care center, bank, shopping center, warehouse, furniture store. The list goes on. If you are an apartment builder, the other apartments in town are your competition.

Let me rework one of the Ten Commandments and introduce: "Love Your Other." Your competition is good for you. Perhaps you've been in a skating rink and thought, "If they would keep better floor rules and improve on their pizza, this place could be a gold mine." You don't know a thing about the skating business, but you made an observation based on needs.

Sometimes our competition is too close to the forest to see the trees, or so bogged down in daily operations that little changes that are needed, and could make a big difference, slip by unnoticed.

Consequently, if you take that same skating rink, and by proper site selection, put it in a nicer area of town, more centrally located, then you have learned from the "other." Of course, you will want to straighten up the floor rules and improve the pizza while you're at it. Why wouldn't you "Love Your Other," if they help you by proving what does and doesn't work.

Analyze the competition—what do they offer, how does it compare to you?

You deal with competition just like you do any challenge. When looking for a site and the competition is there, you either fold your tent and leave, or plan your strategy to win. You decide to outsmart the competition. You check them out and see what they are doing, how it compares to you, how you can be better than they are, and out-do them. This competition stands between you and what you want.

You need to analyze the competition. Why are they there, what do they have to offer, how does it compare to you, can you improve on the situation, and is there room for both of you.

The ''user pies'' in this chapter will show you how to identify your competition, analyze the market, and determine if there is a share of the business you can claim.

THE USER PIE

What in the world is a user pie? No, it isn't a pie full of cocaine or drugs, but a pie full of users. In this book, a user is someone who is going to use your product. Your product can be a brand of paint, a roller coaster, or a piece of real estate on which to build something.

A user pie represents the group of people, or the estimated percentage of the population, who you feel is your customer or client. It could be a percentage of a certain age group, income level, or marital status.

For instance, if you are building apartments that are designed more for the single population, then your user pie is full of the singles population. If the singles population is lower than the number of apartments now available for singles, then your "singles apartment user pie" is all gone. Getting the picture?

On the other hand, if there is a new high-tech company in town that is hiring a lot of young singles just out of college, and all they can find to rent are large, three bedroom apartments designed for families, you've got pie.

The Hamburger User Pie

The hamburger user pie signifies what percentage of the people in the total population are going to eat hamburgers today. The hamburger user pie also signifies how many of these people will eat

Competition is good for you. It makes you work harder and improve where there's room for improvement.

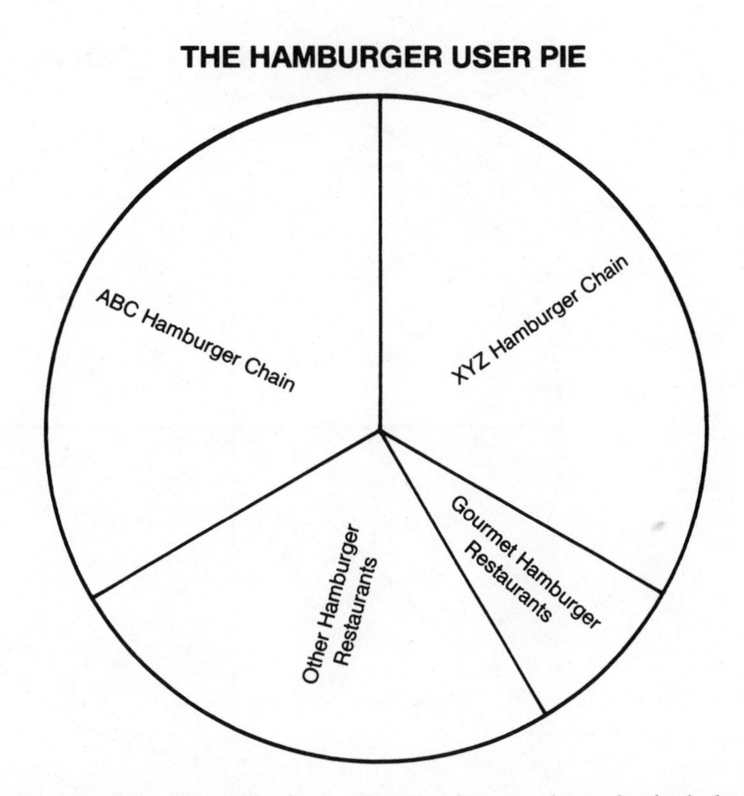

at McDonalds, Wendy's, etc. These pies are hypothetical, however, I am sure McDonalds, as well as other national chains know what their actual percentages are. The actual percentages are based on previous sales along with projections of future sales.

The Restaurant User Pie

Your user (you, your boss, or client) might have a piece of the overall pie and then have to share that piece with the competition. For instance, in the restaurant user pie, each piece of pie represents a specialization. There may be very little or no competition in the pizza restaurant category, but too much competition in the steak restaurant category.

The Department Store User Pie

Let's say for instance, that a shopping center developer has a department store that wants to lease space in his shopping center.

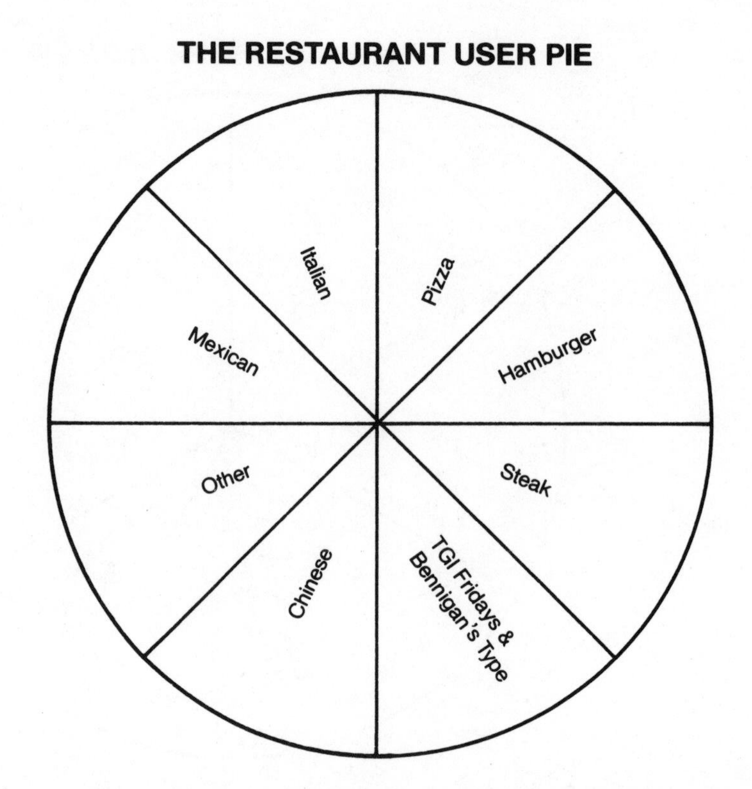

THE RESTAURANT USER PIE

The overall department store user pie represents the total dollars the population in that area will spend in a year on the goods from a department store. There are, however, different types of department stores depending on price, products, etc. The piece of user pie is how many of these people who would spend money at a department store in that market area, will spend at this particular type of department store.

Another way to illustrate this is shown in the department store user pie.

Obviously, if the pie or piece is all gone, you have a problem and *might* need to find another area. If you have *got* to be in this area, it is usually because there is a large piece of pie left and you want it.

So, let's say there is a pie left and you want to be there. The competition is there so what you want to do is do it better.

You want your site to be better. Better access, better visibility, better tenants—which are more tailored to the needs of the area or the customer, better positioned to the traffic flow of the

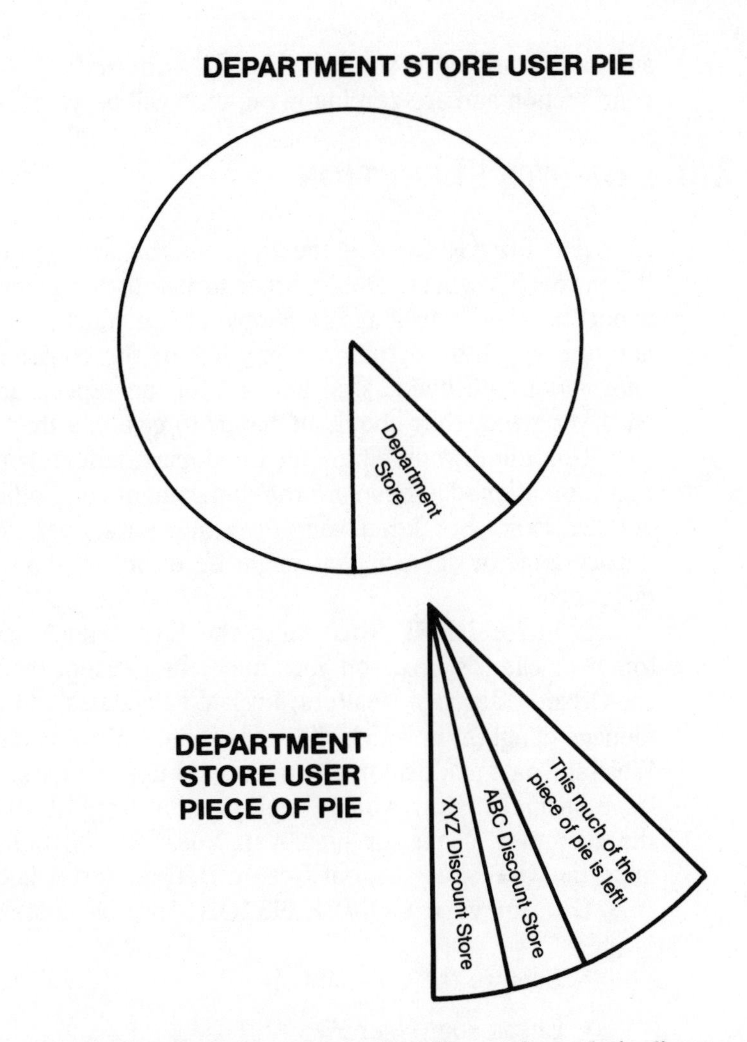

customer—closer to where they live and on their direct route to and from work. Better, Better, Better.

It could even include better topography, such as no trees or fewer trees to increase visibility, or more trees if that is what you need, or perhaps a small lake.

In addition, the way your concept fits the site as well as the price, could play a major factor for you. Maybe you can pass a price savings along to your client or customer. Remember, Better, Better, Better.

In Chapter 13, Fifty Flags to Site Selection, you can review some major factors for site selection. These flags represent red flags that should demand your attention as you are site selecting and particularly after you have identified your city or county, found your section and are zeroing in on what will be your Best site.

THE ABC'S OF SITE SELECTION

A is for AREA: Feel the area, see the area, get in touch with the property owners. Don't forget to talk to the property owners about the area as well as the people at the courthouse. If you are not sure which department at city hall or the courthouse has the information you need, stop into any of the offices and tell them what you need, they should be happy to give you that information.

Don't limit yourself to only the department and other sources I have mentioned. These are the departments and offices you cannot overlook. But keep your eyes and ears open to any other department, or agency, that might be of help to you. Ask lots of questions.

B is for BEST Site: Keep the time crunch and your customer or client's *needs* on your mind. In a recent discussion with the Orlando Board of Realtors, I related a discussion I had with my teenage daughter in which she said to me, ''How would you feel?'' What she said made a lot of sense, and I use the mental exercise a lot in business. How would I feel, or what would I want, if I were the customer, client, or employer. Location, location, location is important in real estate, but Better, Better, Better is critical!

C is for your COMPETITION: Your basic steps are:

- Analyze the Competition
- Check your User Pie
- Choose a site that is the Best

Love your ''other.'' Learn from your competition. Sharpen your skills and expertise against his experience and go for it. Take the *leading* edge. Oh, and call your mom—tell her to go ahead and rent out your old room, you won't need it!

SECTION 3

8

Remodeling and Income Property

If the site you have chosen has a building on it that you plan to remodel, or it is income property you are buying as an investment, remember that this building was built according to zoning and building codes in effect at that time. They could have changed since then.

REMODELING

If you are only going to paint and move, current zoning and building codes should still apply, but check to make sure. On the other hand, if you are planning an extensive remodel, expansion and architectural change, you need to check all of the departments at city hall or the courthouse (see Section 1) and verify that it is possible to do what you are planning. Don't forget the dangers of grandfathered-in zoning.

If you plan to remodel a very old building, check to see if it is classified as historic. A historic building has some very strict remodeling procedures to go by.

When you remodel, do a good job. Make a new look for yourself. Be careful not to botch it, a good site deserves a good face lift.

There are several considerations before buying property for investment other than repainting.

INCOME OR INVESTMENT PROPERTY

When you buy property for investment, the income that has been established is going to be the first and most important thing you will want to see. You are right to verify these facts and to make sure they fit into your investment and tax strategies.

In the process, however, don't overlook changes that might have taken place regarding the site. If the income property you are considering is a shopping center, there could be a new center with very competitive tenants going in across the street. If it is a large apartment complex, the local manufacturing plant could be shutting down, and your tenants will have to relocate.

REMODEL AND INVESTMENT PROPERTY CHECKLIST

Present Building Size ____________ Adequate ____________
Present Number of Parking Spaces ____________ Adequate ____________
Present Size of Site ____________ Adequate ____________
Ingress ____________
Egress ____________
Plumbing ____________
Wiring ____________
Zoning ____________
City ____________ County ____________
Utilities: Water ____________ Line Size Adequate ____________
Gas ____________ Line Size Adequate ____________
Sewer ____________ Line Size Adequate ____________
Septic Tank ____________ Location ____________
Storm Water ____________
Other ____________
Easements ____________
Restrictions ____________
Road System ____________
Parking Code ____________
Landscaping Code ____________
Why Previous Occupant Left ____________
Positive Features ____________
____
Negative Features ____________
____
Competition: Present ____________
Proposed ____________
Other ____________
____

Fig. 8-2. *Use this remodeling and investment checklist before deciding on a piece of property.*

Check everything out; go to the city hall, the courthouse, and the chamber of commerce and get a handle on what is going on in the community around this site. A new bypass could cut you off from traffic flow. Know what you are buying.

The Remodel and Investment Property Checklist shown in Fig. 8-2 can be used as a guide. Checking all these things out really doesn't take that long. In most cases, one day at the city hall or courthouse, and buzzing around town should tell you what you want to know. Don't rush, however, it could cause you to forget something important.

9

Site Selection in a Tourist Area

Most of America vacations as a family, but keep in mind that many of today's travelers are singles, mingles, and empty nesters. You are really going to like this chapter because in order to analyze the situation, you will just have to join the fun.

There are basically two ways to approach having fun. Start a brand new resort area, or get in on the action of an established area.

GETTING IN ON THE ACTION

To start a whole new vacation area or tourist attraction, first you need to decide what type of playground you want to develop, who your market is, and how to get there. This book is not about how to make this decision, but how to find your site after you've made your decision. Be sure to do your research well so you will have a successful project.

To get in on the action of an established area, it is critical to watch your Timing. You will be successful with your site when you supply what is needed at the time. I've seen great ideas or projects absolutely fall on their face because it was ahead of the demand (needs), or after the supply had been met. Sometimes one project will feed another, but the Timing must still be right.

Before starting a tourist-based business, determine who your market is—families, singles, retirees.

TIMING

Not only is Timing important in site selection in a tourist area, but in all phases of site selection.

Here we go again with something else to check out. Not only do you have to check supply, demand, but now Timing. Timing is so important that I probably should have made Chapter 1 about Timing. I always like to express Timing with a capital "T" in order to give it just some of the respect it deserves.

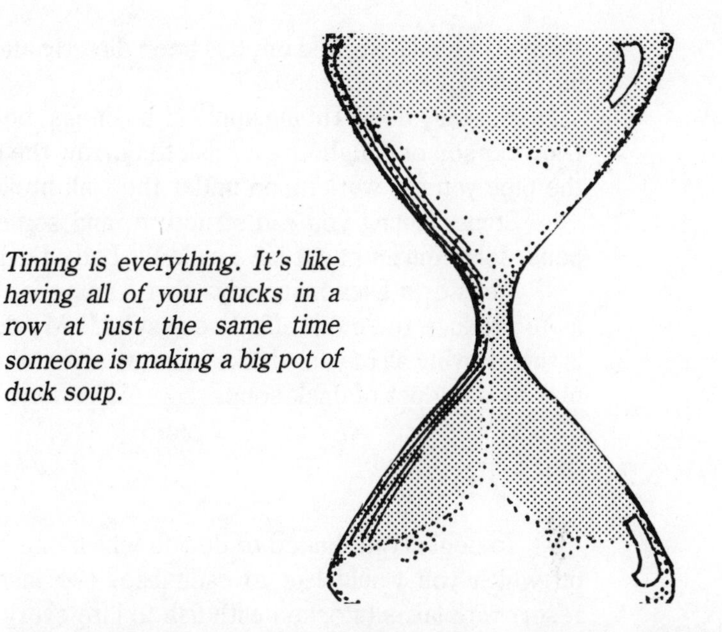

Timing is everything. It's like having all of your ducks in a row at just the same time someone is making a big pot of duck soup.

Timing is everything. If it is hard to remember this phrase or cliché, maybe you should consider printing it on a card and placing it where you will see it everyday. Maybe on your bathroom mirror where you can see it when you shave or put on your makeup, or on your desk at the office. I suggest the bathroom mirror because it not only applies to business decisions, but Timing can help with your personal objectives also.

I can't take of all of the credit for figuring this out. I worked on a project a few years back with Lloyd Smith, a very successful real estate investor from Bloomfield Hills, Michigan. Lloyd had come to Florida in the days before Disney, and again in the early seventies when development had hit a curve and some very good land buys could be had. Over the years, Lloyd turned these good buys into profits of millions of dollars by using the right Timing.

Lloyd and I worked on a project for almost three years. As I remember, Lloyd repeated the phrase "Timing is Everything" to me every day for three solid years. Sometimes I got tired of hearing it, but I later came to realize how true it was.

I think these words are imprinted now somewhere in one of the hemispheres of my brain, because it has become a natural instinct when working property to always consider the Timing. I am convinced that the project Lloyd and I completed, and everything else I have worked on, has been directly affected by considering Timing.

Not only does Timing apply to business, but personal as well. Did your son or daughter ever ask to borrow the car about the time the pipe you are working on under the sink broke loose again?

Some Timing you can structure, and sometimes it just happens. John Wayne might have called it Lady Luck.

Webster's Dictionary says that Timing is, "of something so as to produce the most effective results." My definition of Timing is this: Having all of your ducks in a row at just the time someone is making a big pot of duck soup.

THE NEW RESORT

To begin, you'll need to decide which one of God's treasures on which you would like to capitalize. For instance, a mountain resort with lakes (stocked with fish to lure every fisherman far and wide), streams and nature trails. This summertime fisherman's paradise could also be turned into a snow skiing resort in the winter. You could have discovered a natural wonder—comparable to the Mammoth Cave in Kentucky or the Redwood Trees in California; or develop a tennis and golf resort in an area where the weather is so perfect you can play tennis (or golf) 365 days a year.

Let's say you've picked your project. Sometimes your project itself will dictate what general area you must be in. Keep in mind that, depending on the type of project you are considering, you could look for the area you want to develop on a national level as well as a state or city. The point is, you really shouldn't look for property for a snow skiing resort in Florida, or 80-degree temperatures in January in Ohio.

You really have your work cut out for yourself. As I previously mentioned, depending on the type of resort you will be doing, you need to pick an area that is geographically and climatically suitable. Once you have done this, you can pick smaller areas within the larger one to zero-in on. For instance, if you have decided on a particular state, then you can locate the areas within the state that are suitable to your project.

When you have narrowed your location or locations down, and have determined the city or county (maybe even state—depending on the type and size of your project) municipality which will govern the development. Then you guessed it, go back to Chapter 1 and go through the process of finding an area within that geographical or climatic perimeter which, according to the Comprehensive Land Use Plan, can be used.

It is very likely you will have more than one location in which to look. For example, if you have decided on the beautiful Blue Grass State of Kentucky for your Dude Ranch, there might be three towns that have adequate airports and road systems to accommodate your clients or "dudes." The time crunch comes into play here too. People are also limited for time when they are on vacation, so they will want to be able to get to you with a reasonable amount of convenience in a reasonable amount of time.

If your resort is large enough, you might be able to get your property zoned as a Planned Project or Planned Unit Development by providing your own utilities, road systems and fire protection. As mentioned in Chapter 1, one of the main reasons for a Comprehensive Land Use Plan and proper zoning rules and regulations is to control and plan the growth of an area and enable the municipalities to plan ahead in an attempt to provide adequate utilities, roads, fire and police protection. I say attempt here because sometimes rapid growth makes it very difficult.

Remember the needs discussed in Chapter 6. If you are planning a ski resort, you are going to need a mountain. This doesn't really appear to be one of those characteristics you can make if there isn't one there.

You can make a lake, add a hill, or provide beautiful landscaping, but be sure to recognize that everything you add costs money. And, these costs must be added to your feasibility study. Try to pick the piece of property that best fits your needs to keep these extra development costs to a minimum.

Go to the competitive resorts or tourist areas and check in. Stay a few days, participate in the fun (I told you this chapter would be enjoyable) and see what you could add or delete to be better. Remember, you want to be Better, Better, Better.

Take plenty of time and get all the professionals necessary for your project. As Mom said, "Do Your Homework."

AN ESTABLISHED RESORT AREA

You saw a picture of a tourist at the beginning of this section. Don't let me scare you to death, you don't have to look like that to find a site in a tourist area, however, you will need to jump in and "be one."

Let's pack our play clothes. We're going to the action. The reason it is important to stay in the tourist area a few days and check it out is because these areas can be the trickiest of all. If you are reluctant to agree, I could get you the names of some investors and developers who were in Florida in the early seventies.

Disney World had "come to town" and there appeared to be "no way you could go wrong." To people who had seen Anaheim, California develop and what Disneyland had done for it (or to it some might say), it was a sure thing for Central Florida.

Walt Disney's purchase of thousands of acres in Orange and Osceola counties in the early seventies is to me, the very best example of developing a new resort. To "Get In On The Action," investors and developers from all over the world flocked to the Central Florida area when the announcements were made. Some of the developers were overanxious and had their projects completed way before Disney World was open for business. (Too bad, Fifty Flags to Site Selection wasn't available to them—Chapter 13.)

Disney World had not been opened long, when overdevelopment hit hard. From too many apartments for the workers, to too many gift shops with souvenirs, and everything in between such as motels and restaurants. There was just more supply than demand.

If you want to add to a resort area already established, you'll want to be extra careful with ole "Mr. Supply and Mrs. Demand," and their ever-loving offspring "Master Timing." Keep an eye on this baby.

APPLYING THE BASIC ABC'S

Check out your market—the type of people who will come to this resort. Are they mostly families, business convention attendees, singles or retirees? If you want to have a lasting impression of the differences in resorts, take your family to a peaceful, relaxing vacation in Daytona Beach, Florida during Spring Break.

Stay in the tourist area for a few days, weeks or whatever it takes for you to feel like you know what is going on. Talk to the people—other tourists, managers, waitresses, and townspeople. It is well worth the time and money. It is critical for you to "do your homework."

Back to the courthouse of course, and find out all the information about development in the tourist zoned areas. An example of the zoning is shown in Appendix A. This zoning will have all of the requirements such as drainage, parking and setbacks as we have mentioned. Go through all the processes. Find out where zoning could be available if there is no property presently zoned for what you need. Ask lots of questions. Go back to the ABC's:

A Find your *Area*.
B Find the Site that *Best* fits your needs.
C Check out the *Competition*.

When you have found your site and checked out all the development criteria, you are ready to make a decision, hire professionals and proceed. Don't forget to re-check the Fifty Flags to Site Selection.

Good Luck with your tourist area project and, don't forget to have fun.

10

Branching Out

Often, when a business branches out it can seem more like "Going Out On A Limb." You have probably seen this happen. A dry cleaners, restaurant, or business office in town is doing great so the owners decide to expand by opening another store or office.

They think that because they are established and know how to run their business in one area, they can set up in a new area and business will automatically come to them. So that is what they do.

Then disaster hits. The new place doesn't do well and the original store which is still going great has to start feeding it. Pretty soon they are both closed and everyone wonders what happened.

What did happen? The careful site selection followed in the opening of the first unit went out of the window when the glitter of expansion hit the horizon. The "I'm the expert, I'm successful. Hey—no problem," syndrome set in.

Perhaps you've seen this entrepreneur? He brags to his friends, "Oh, yes, we're doing 1,200 customers a day here. I built the business from scratch, people go out of their way to do business with me."

Well, open up in a bad location and see how far out of their way people will go, and how often. You can keep this from happening with just a few simple steps that you've already learned.

If you're planning to branch out, consider what makes your first site so successful and use it as a blueprint.

Perhaps the success of your first store was absolutely luck. You took it over with nothing down because the owner was broke, or in bad health and, with a few minor changes, some ideas of your own, the business flourished and put you in great financial shape. Wow—what a beginning.

If you have been fortunate enough for this to happen to you, back up a step or two. Let's determine what happened. What does your current site have going for it that you want to find in another site. Opening up a second store or expanding your market should not be a "Hold your nose, close your eyes and jump," experience.

If you have a good thing going for you, use it as the blueprint for your next location. Use the site questionnaire in the following section as a guideline to choosing your new site. If you want to do it right again, you need to find out what you did.

SITE QUESTIONNAIRE

Carefully analyzing your current business is the first step to successfully branching out. The following questions will help you to think through the elements of what does, and does not, make for a winner.

Customer

Who is your customer? Remember the user pies? It is very important for you to identify your customer (your User). Establish your customer's age, income level, race, family status, etc., and even how many of these customers it takes in a market area to give you the sales volume you need. See if the population has been growing or declining during the time you have been established and how this has affected your sales.

Destination Point

Is your present location at a Destination Point? If it is a Destination Point itself, determine what made it such a success. Is it the tenant mix, the size of your User Pie or ''piece'' of User Pie, the road system, the visibility and accessibility? If you are at a Destination Point, what has made the Destination Point successful? Also, how does your site fit in with the rest of the area? For example, are you at an entrance to a shopping center, are you visible from the main road, are you on the ''Going Home Side?''

Road System

Analyzing the road system is very important. We have talked much about roads. The time crunch is a reminder of the fast lane we all live in, so be aware of any changes in the present road system such as a new bypass, or plans to widen or resurface the present road. A new bypass could cut off your traffic flow and widening or resurfacing the road or street in front of you, can certainly kill your business.

Ingress and Egress

Ingress and Egress of your site, as well as the safety of your accessibility is very important. Check to see if you are on the Going Home Side or the Morning Side and, if so, how you think it affects your particular business. Make a trip to the department of transportation to get an accurate traffic count of the street or road on which you are located.

Size

How does the size or configuration of your present site work for (or against) you? Could you use more parking? Is the site long and narrow, causing your customers to walk a long distance to get to your door? Is the flow of traffic on your site good?

Finally, is the size of your building adequate? Could you use more seats, more room in the kitchen, or more space for merchandise, office, or display?

Physical Features

Are there physical features you will not want to live without at another location? A lake used for boat sales or a seafood restaurant, or trees for a residential development? Perhaps there are physical features you definitely don't want again—a small hill right at your entrance?

A Generator

If you are a generator, what has helped you generate the most people and sales? If you are close to a generator, or a group of generators, study them and look for a site that you can duplicate what you have. Generators make the Destination Point. Consequently, a popular and varied group of generators can make you very successful.

Visibility

Pretend to be your customer. Drive up and down the street and check your visibility—what is good about it and what isn't. Check on:

1. How is the neighborhood?
2. Where is the new growth?
3. How is the User Pie—and all the pieces?
4. How is your Timing? Anyone making duck soup?
5. What needs have you supplied?
6. Is there something you need at your present location?

7. What are your cost parameters to keep prices the same?

8. How does your feasibility study look?

Getting the idea? You must know what I am going to say next. Make a trip to the courthouse and/or city hall. Go through the process as outlined in Sections 1, 2, and 3 (the ABC's) and review the 50 Flags. Give the second (or third or fourth) location all the advantages of the first one that has done so well for you. Maybe you can even make it Better.

11

Relocating Your Headquarters

There is a lot of company relocation activity today. Many decade-old companies are looking for new locations. Sometimes a change in operations and products, makes relocation not only possible, but necessary. Manufacturing industries as well as high tech and aerospace companies are just some industries that often require a new location.

Stiff competition, growth, and specialization can also prompt a company to look for a more attractive location. In the last 10 years, for example, Florida has become a very attractive state for company relocation, branch plants, offices, or new companies. Florida is fast becoming an economic and industrial leader in the Southeast. Lower living and operating costs, no state income taxes, a nonunion environment, ample waterways, railways, international airports, flexibility, diversity and, last but not least, a warm sunny climate, are among the top reasons Florida has experienced such an economic boom.

ANALYZING YOUR HEADQUARTERS NEEDS

Your company may have other needs that more or less limit where you can locate your operations. A coal mining company needs coal; a logging company needs trees, etc. If you have a company that is being swallowed by overhead, or is busting at the seams, however, other criteria will be a consideration.

After you have chosen a prospective state, you have to decide on a community that can provide you with what you need to operate successfully. Then the site selection begins.

There are major research companies whose expertise is to match your company's needs with the right location. A complete profile is developed and presented to you for your company's consideration. This chapter is a basic guide to help you get started.

CHOOSING THE STATE

This is going to sound a lot like Chapter 6 on Needs discussed in Section 2. Before selecting a state, a company must outline their needs. Be sure to consider the labor force, unions, taxes, state government, population, growth trends, business climate and quality of life. Facilities and skills together with accessibility to the market place for distribution, is a must.

A company will need to find the right combination of location and labor force, together with a business climate that will promote growth.

In addition, a relocation company could need a railway spur, waterway accessibility, an international airport, major road systems and even a particular type of climate.

Other important considerations are:

1. Quality of education for employees as well as their children.
2. Cost and availability of quality housing.
3. Utility availability.
4. Specific skills.
5. Product or raw material availability.
6. Network possibilities.
7. The state's acceptability of the company.
8. Proximity to market or customer.

9. Political Structure.

10. Finding an acceptable community within the state.

All of the negatives and positives should be evaluated carefully, taking into consideration growth trends, similar industries currently located within the state, and the distance from the current location. Relocation costs and advantages should always first be weighed against the costs and advantages of expanding and improving the present location.

CHOOSING THE COMMUNITY

Once you have found a state that fits all your company's needs, the next step is to find the right community to complete the picture.

The community location is important. The community should be conveniently located to the road system, airport, rail and waterways, that you need. Also, you will want a community where a large percentage of the type of labor you require lives.

The local chamber of commerce is a good place to start to learn more about the community. The director of the chamber of commerce can tell you a great deal about the community, as well as provide you with the names of the community leaders. The mayor, industrial development director, chief of police, airport director and even the superintendent of schools are a few of the people you can talk with about the community.

The same criteria you looked for in a state also applies to selecting a community. The quality of life for executives and other employees is important for a content labor force. This has been established as a must in reducing employee turnover.

The quality of life can include the availability and quality of medical facilities; the crime rate; police and fire protection; and the quality of schools, for all ages; and recreational facilities.

The price of quality life is also very important to your employee and his family. A quality of life that is not affordable to your employee can cause you real trouble in unexpected and necessary wage increases. The cost of property, the local real estate tax structure and insurance rates play another important part in community affordability.

Find out if there are any big changes planned for the community. For example, a new state jail or mental ward, or a new race track could promote a negative environment. Also, look for upcoming positive changes such as a new school or interstate.

Talk to other industries and companies in town. The chamber of commerce has information on these companies, but also check with companies who might not be listed with the chamber of commerce.

Up to this point, you have been very careful to consider your employee. Now it is time to talk about you—the company.

Is the community interested in your company? If your company is a manufacturing company or an industry, find out right away if the community wants that type of development.

The employees will be happy and the community is romancing you for marriage. What is next?

Does the community have the site or facility your company needs?

CHOOSING THE SITE

No matter what area of the community you selected for your site, it will be governed by local zoning or the Comprehensive Land Use Plan.

Beginning with the courthouse as we did in Section 1, is exactly what you need to do. The zoning and Comprehensive Land Use maps tell you where in the community to look for sites or facilities.

Follow the procedures in Section 1 and Section 2 for your site selection. If the competition is a factor in the selection of a site for your plant or headquarters, Section 3 can help.

If you haven't read all of Sections 1, 2, and 3, you need to do so now so you can have a basic guide for finding a site in your community. Don't forget to review the 50 red flags in Chapter 13.

SECTION
4

12

The Feasibility Study

There are not very many people who want to do a project to lose money. The bottom line is: ''Will it make money, how much, and when?'' This sounds like three bottom lines but it really isn't. Money spent (invested), and money received (return *of* your investment including return *on* your investment), balanced against the period of time passed, results in the bottom line.

There are companies that can prepare a feasibility study for you. A Certified Commercial Investment Member (CCIM) of the Realtors National Marketing Institute, accountant or engineer might also be able to prepare a feasibility study for you.

A professionally prepared feasibility study is very helpful, and most of the time is essential to present to your lender if you want to obtain a loan.

MONEY SPENT

I have used very simple terminology for this section. Down payment, initial investment, investment, up-front money, capital, and project costs are some other ways of putting it. Money Spent should be a realistic, complete estimate of what your project will cost. Get all the facts and figures. Leaving something out can be

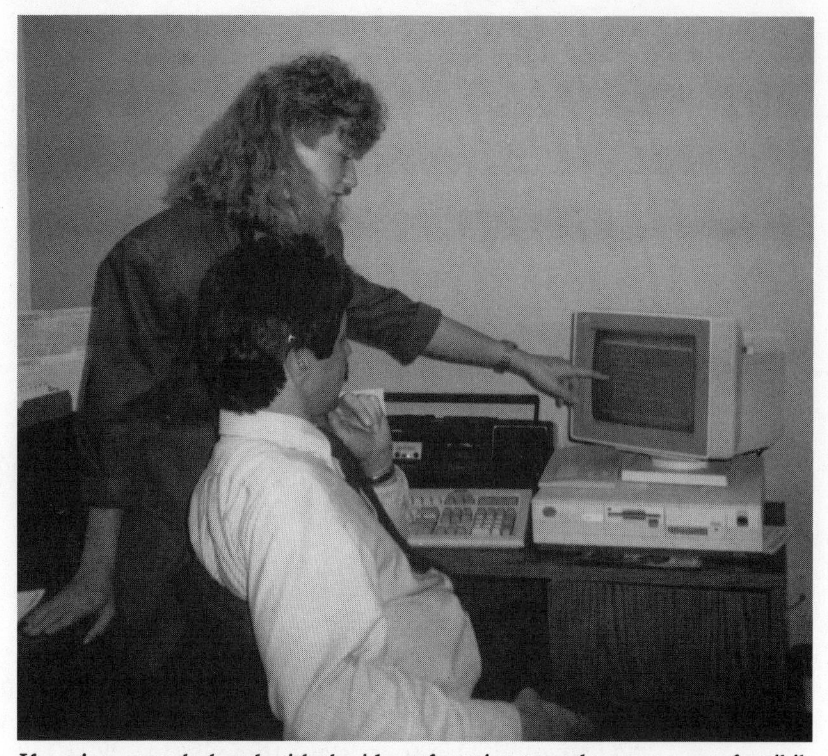

If you're overwhelmed with the idea of putting together your own feasibility study, have a professional do it.

disastrous. An omission could cost you your job, cause you to run out of money and be unable to complete your project, or relieve you of your life's savings. The following are some expenses and costs to include in your feasibility study. Your project might include other costs and these should be factored in. Always have a professional help you.

Accounting Fees

You will need an accountant to assist you with financial statements, feasibility studies and to plan your tax structure.

Advertising

You will need advertising, marketing and promotions for your project.

Architect Fees

Building plans and site plans are needed by an architect certified by the state and county your project is in. These are needed for new developments, as well as remodeling present structures.

Attorney Fees

Preparing your purchase contract, reviewing options, title work and estate planning are a few of the services you need from an attorney.

Building Contractor

You need someone to apply for permits and build your buildings. This is your guy.

Engineering Fees

As discussed in Chapter 14, Professional Services, you need engineering for your site work, drainage, roads and utilities.

Impact Fees

The costs to hook up to the utility systems, costs of increasing traffic on the road systems and other fees by state and local agencies.

Taxes

The income tax charge or savings generated by the project. Also included are real estate taxes, personal property taxes and state and local sales tax and income tax. Your accountant or financial planner can help with these.

Insurance

Depending on your project, you could need insurance coverage such as builder's risk (covers property under construction), worker's compensation (for your employees should they get hurt on the job), fire and extended coverage on the completed structures, liability (in case someone gets hurt on your property), and possibly group life and health insurance. Your insurance agent can advise you on the type of coverage you will need and the premium costs involved.

Interest

Interest should be added to your feasibility study, whether you are paying it, or have paid cash and are losing interest. Include interest incurred while under construction, as well as interest on your permanent loan. Don't forget to include those elusive points and loan origination fees.

Land Cost

The purchase price of your property. This could also include permitting fees and real estate taxes.

Landscaping

Required and/or desired trees, plants, shrubs, grading, etc. necessary for your project.

Licensing Fee and Permits

These fees vary in each state, county and city—make sure you get them *all*.

Maintenance

Maintenance of equipment or portions of the completed project could be needed. Security might also be a requirement. A night watchman could be necessary to reduce theft of equipment and materials.

Marketing

Promotions and marketing for your project can give you a big boost in recognition. A well-planned and advertised grand opening can get you off to a great start.

Materials

This list could be never-ending. From nails to petunias, you'll need materials for the job.

Operating Costs

Salaries, benefits, utilities, telephones, office rent, supplies, and a host of other expenses necessary to get, and keep your project going can really add up.

Roads and Utility Lines

Upgrading present roads and/or utility lines. Installing new roads and/or utility lines cost money. You won't always have to do either of these but if you do—add them in.

Site Work

Watch out for this one. Site work has a talent for surprises. Make sure your estimates are complete and that your soil test is complete.

Traveling Expenses

Sometimes this one is overlooked. The traveling necessary to put a project together can quickly mount up.

Make a checklist for yourself using these as a starting point, adding other costs, fees or expenses relative to your project.

MONEY RECEIVED

Again, "money received" is a simple term that refers to income, cash flow, gain, return of your investment and return on your investment. Return *of* investment is getting back the money you put in. Return *on* investment is what money you got back over and above what you put in. For instance, if you put $20,000 into a project and received $30,000 back, the $20,000 is return *of* investment; $10,000 is return *on* your investment.

While there is no magic number or percentage, you should *always* be able to make a profit. Each project is different. Usually, if the returns are greater, so are the risks.

What are risks? If you go to your favorite bank and buy a certificate of deposit, you are guaranteed two things: A return *of* your investment and a percentage return *on* your investment—no risk, because it is guaranteed by the bank.

On the other hand, if your local broker brings you a package on a piece of real estate showing that you can buy it and resell it in one year at a 25 percent return *on* your investment—this baby has risk. No guarantees but plenty of potential.

You can reduce the site selection risk by going through the ABC's outlined in Sections 1, 2, and 3 and the 50 Flags to Site Selection to check the facts and eliminate surprises.

Cash Flow

If you are an employee, cash flow is your salary; if you are a dress shop, it's your sales; if you are an amusement park, it's your ticket sales; if you are a teenager, it's your allowance. It seems as if cash flow is what makes the world go ''round.''

Cash flow has a personality like a mountain stream: it can go up, down, freeze up, or come to a complete stop. For feasibility purposes, cash flow can be considered ''before expenses, after expenses and before or after taxes.''

Gain

A ''gain'' is a term used in reference to the sale of a project. If you plan a sale at completion of your project, the gain will be addressed as part of the return *of* and *on* your investment. Using the previous example of the real estate you can buy for $20,000 and sell for $30,000, the $10,000 could be referred to as your gain. As with cash flow, you can have a gross or net gain, before or after expenses, as well as a before or after tax gain.

TIME

The time value of money should never be ignored. If you are to invest in a project that will not be completed for five years, you have to include the cost of money in your feasibility study. The cost of money is most simply understood as interest. If you borrow money of course, you will be required to pay interest on it. But be *sure* to recognize that if you put cash of your own into a project, you must include the *loss of* interest on this money while it is in the project. So be sure to address the interest on borrowed money or the *loss* of interest on cash.

The more time it takes for a project to generate Income or Gain, the more risk is involved. Like that ol' mountain stream, you don't know what is around the corner. There might be a large waterfall or a couple of energetic beavers building a dam for the Guinness Book of Records.

Although you cannot see into the future, by checking the demographics and growth development patterns already in motion, you can get a good picture of how the area and site you have chosen should develop and progress.

13

Fifty Flags to Site Selection

Flags here mean *red* flags—things you need to watch out for in site selection. Some, if not any one of these, can be site killers.

Believe it or not, I have run into every one of these problems in my work as a commercial real estate broker and site selector for a national franchisor. These are just *some* of the problems that can hit you between the eyes. Be careful and check everything. Don't rely on hearsay or birds-eye viewpoints—get the facts instead. The professionals you hire in Chapter 14 should be able to help you determine the facts.

1 PROJECTS COMING SOON

It is very easy to put up a sign that says, ''Project Coming Soon.'' The intentions are usually good, but sometimes for one reason or another the project never gets off the ground. If you are looking for a shopping center site because a large residential development and office park complex has been announced, it is important to arrange the purchase of your site in such a way as to allow yourself time to check the validity of the announced project. This is done by checking the principals of the project for financial strength and the success of other projects they have done. Go to the courthouse for this information and check on the project's permitting to see how far along the project is.

Permits are needed on any project. The larger the project, the more permits it needs. A large residential project needs permits from such agencies as the department of environmental protection, the department of natural resources, the water districts, the building, zoning, and planning departments and even a host of other approvals for the engineering, roads, and utilities. A project with all these permits has a lot going for it, and chances are, if the principals involved are not able to finish the project, it can be sold to someone who can.

2 INCOMPLETED PROJECTS

When the projects do actually begin construction, it is not unusual for a project to hit a snag, run over budget, or simply be poorly planned and come to a dead standstill before completion. These delays can destroy your project or add holding costs to your site that destroy the feasibility. For example, buying an out-parcel to a shopping center that is not completed or remains unleased.

3 TRAFFIC JAMMERS

Traffic can be wonderful and you sure want all you can get. Watch out, however, for the site that is locked in by traffic. Sometimes a corner can be this way, especially for a smaller user. A corner with a lot of traffic may be easy to get into by making a right-hand turn into, but a brick wall of backed-up traffic waiting for the traffic signal can make it impossible to get out of, or to make a left-hand turn into, the site.

4 INGRESS AND EGRESS (GETTING IN AND OUT)

Your clients or customers will not want to endanger themselves or their families getting in and out of your place of business. Neither will they want to have to call a traffic engineer to determine how to get to you. Also, if your use is industrial or manufacturing, a difficult entrance and/or exit could cost you thousands of dollars in missed or late deliveries.

5 TREES

Trees are beautiful, there is no doubt about it. But, when you find a site that is dense with trees, has unusually large older trees such as oaks and maples (or perhaps has trees on it that have become *endangered*), this can very likely mean trouble. More and more areas are on a "save the trees" campaign, and you could find after checking with the engineering department that you cannot remove some of these trees. You will need to obtain a permit to remove the trees and also provide the engineering department with your site plan showing which trees you want removed and which will stay. Sometimes you can do some tree swapping. For example, you can plant some trees in the city park or replant on your site in more strategic places as an exchange. Don't try to slip in there in the dead of night and remove the trees before you are told you cannot. I have heard of developers who tried that and were heavily fined, and what is worse, their project was put on hold for an extended period of time.

6 ENDANGERED TREES OR PLANTS

Yes, trees can also be endangered species. Different parts of the country have different types of special trees. I call them special because they have become so few in number. When you are looking at property that has an unusual tree or especially a *group* of unusual trees you had better check it out.

When a developer I worked with in South Florida tried to clear a grove of mango trees to put up a strip shopping center, the city said "no dice." This, by the way, was after the closing on the property. The mango grove had to stay. So if you are ever in South Florida and notice a nice mango grove in the middle of the city where a grocery store would go nicely, you'll see what I mean.

Some plants and flowers are also endangered. If you suspect there might be an unusual plant, tree, or flower on the site, call a horticulturist for an evaluation, before closing the deal.

7 MORE TREES

When you see a site with a lot of trees and foliage, your red flag should say how much is it going to *cost* to clear this site if I *do*

every easement, every setback and can wind up with a project that doesn't have adequate parking or adequate retention, etc.

If the piece of property is too small, either find a piece that is large enough or reduce your project to fit the property. Don't try to squeeze a project on a site that is too small for it.

10 SIGNS

Every city and county has a sign ordinance or a set of rules that you have to go by for your sign. The sign ordinance governs the size, height, placement and sometimes even the color of your sign. Hilton Head, South Carolina is a good example. All signs must be small, low to the ground, and billboards are prohibited. It is something you will want to address and analyze. It does help when your competition has to conform to the same ordinance, but it will probably make you somewhat envious (and nervous) if your competition was there before the sign ordinance was put into effect and has a big, beautiful sign that you can see from everywhere and *your* sign falls under the new sign ordinance requirements.

11 COMPETITION

We have talked about competition and we know from the old adage that it is good for the soul. However, if your competition has you out-positioned from day one and you will always be on a site secondary to theirs, you will want to consider this very carefully. If you just *have* to be there and you want to take a chance on this site, try to negotiate some extras with your site that could help. For example, try to negotiate the price to reflect the possibility of doing a lower volume of sales.

12 DON'T GET CAUGHT BEHIND ANOTHER BUILDING

If you are a restaurant or a bank and you are on an out-parcel of a shopping center, check the site plan for the entire project to make sure there are no surprises. Make sure there isn't a site designed to fit between you and the road that would affect not only your visibility, but possibly your access.

Beware of sites that might have large, older trees.

get a permit? The cost of clearing a site can be a real unpleas surprise. Also, when a site needs a lot of tree clearing, it usua needs some fill dirt. I'm not saying you can't do it, just make su you can before you buy the property *and* include the cost in yo feasibility study.

8 TOPOGRAPHY

Fill, grading, removing boulders, de-mucking and any oth preparation needed on the site can be a nightmare of costs. If yo site needs any of this, get a cost estimate.

9 SITE TOO SMALL

One of the areas in Site Selection that can be absolutely fa for your project is to try to squeeze a project on a piece of proper that is too small. You will have to fight for every parking spac

This can also be true if you are going inside the shopping center. Find out where the out-parcels are planned and try to position yourself in the shopping center so you will not be directly behind one of them.

13 CROSS PARKING

A cross parking easement is the right to park on someone else's property as designated. For instance, if you are going to build a bank on a shopping center out-parcel, and the out-parcel is not large enough to accommodate all your parking, the shopping center developer could grant you cross parking rights to the parking spaces of the shopping center. Zoning will almost always require that you satisfy a certain amount of parking on your own property, and allow you to cross park for the rest.

There are three important factors you will want to check:

1. In some cases, the shopping center does not have parking spaces to give you because they are using all of them to meet their zoning requirements.
2. Check the parking area you plan to use to determine if there actually are any spaces available for your customer. There might be spaces left over according to the zoning ordinance, but because business is so good at the shopping center, there *actually* are no spaces for your customers to use.
3. Make sure the cross parking is convenient to your customer—don't make him have to cross a main thoroughfare or swim a retention pond to get to you.

14 CROSS PARKING, PART II

It is great if you can obtain cross parking with a shopping center or an adjoining piece of property, but you do not want a *reciprocal* cross parking easement. For example, you are on an out-parcel to a shopping center with a reciprocal cross parking easement (you can park with the center and the center can park on your property) and a large health spa rents space in the center. This health club really uses up the parking and, more than likely, it uses it at the peak times of the day and night for your business. These people could park all over your parking lot and you would not be able to stop them because of the reciprocal cross easement.

15 DON'T GET CAUGHT IN THE MIDDLE

Let's say that you have found a beautiful piece of property in the middle of a large residential area and you want to put in a bank, restaurant, or even a small strip center. You have checked and there is a nice shopping center (Destination Point) two miles up the street and another nice shopping center (Destination Point) two miles down the street. You decide to position yourself right in the middle, between these two shopping centers in full view of the 60,000 cars per day that travel this street. Remember the Time Crunch and the Destination Point? These people are very likely to go from one Destination Point to another, making an extra stop at your place of business will cost them time and will be inconvenient. Don't confuse this, however, with a large single-purpose user. If you are a single-purpose user such as a large furniture outlet, this changes the picture because a large furniture outlet is a Destination Point.

16 MEDIAN CUT

Breaks in the median strip of a highway make it much easier and safer for your customers to get to you, or for your truck drivers to get in and go out again. If you are caught in between median cuts, this means your customer or employee must go to the next median break and double back. The best situation is to line up your driveway with a break in the median.

17 FUTURE ROADS

Check future road plans as well as immediate plans for road construction. This can tell you two things: (1) Your business will be temporarily hurt by construction if the road in front of you is scheduled to be widened or resurfaced; and (2) if there is a new road planned, the road you will be facing could become obsolete or secondary.

18 FLOOD PLAIN, GREENBELT, SWAMP AND WETLANDS

A piece of property located in a flood plain, greenbelt, swamp, or wetlands can have serious development problems and the net usable area could, in most cases, be drastically different from the total area or acreage. Check these areas very carefully by contacting the appropriate governing agencies such as the water district or department of natural resources.

19 BAD SIGNAGE

You do not want to be a "secret agent." People must know where you are. If your project is a shopping center, you might get caught under a very strict sign ordinance and your tenants will not be interested in leasing from you. Also, if you are on an out-parcel to a shopping center, there might not be signage left for you because the shopping center, tenants, and/or other out-parcel users have taken it all. Finally, any signage space left might be inadequate.

20 BEING OUT-POSITIONED

This is particularly critical for a small market. If your competition clearly has an A site and you are going to be forced to take a B, think long and hard before you jump on this site—it is a very big chance to take. For example, let's take a town of 70,000, put in a shopping center with four out-parcels. A bank, a tire store, a hamburger chain restaurant, and a chicken chain restaurant. (This is not a chain made of chickens, but a national franchise restaurant whose primary menu is chicken). You think the site that is across the street from this shopping center is dynamite for your new restaurant that specializes in fried chicken, but stop and think about it a minute. If you are at the shopping center (keep the Time crunch in mind) and you want chicken, are you going to cross the four-lane highway to get chicken or are you going to move your car a mere 300 feet, or walk to the chicken chain restaurant at the shopping center?

Better ingress and egress, better visibility, better parking, etc. (here we are back to Better, Better, Better) are also examples of being out-positioned or your competition having a better site.

21 OUTSIDE THE HUB

Don't venture off the beaten path, or too far away from the Destination Point. When you analyze an area, you will see that people stay in an activity area.

You could be in serious trouble if you are *close* to the activity area, but inconveniently located on the outside of it. If you are going to create a Destination Point, check the road system and the travel patterns of your clients or customers to make sure you are in their path, and that they don't take another road just before they get to you. (As mentioned in Flag 17, check any proposed roads. The department of transportation could be planning a bypass that would turn your project into a ghost town.)

22 SEWER CAPACITY

Make very sure you have a sewer for your project. Sometimes there is a problem with the other utilities such as water, gas or even power lines, but sewer capacity seems to be the biggest problem today. You might have to pay impact fees and sewer allocation fees in advance, but it is money well spent.

23 ADEQUATE UTILITY LINES

Make sure all the utility lines are at your site and are of adequate size for your project. Although there are sewer, water, or gas lines, to your project, they might need to be replaced because of new county or city specifications. This would considerably add to the expense of your project. Surprises are not fun unless they are good ones.

24 IMPACT FEES

Impact fees are measured by how much water, sewer or gas (utilities) you will use. There are also road impact fees in some areas, which are determined by how much your project will affect traffic. Other impact fees, such as school impact fees and police and fire protection, can be required depending on the municipality in which you are developing your project. These fees can be quite expensive.

25 SOIL TESTS

Never, never buy a piece of property without obtaining a soil test. Your own soil test. If the seller has had the soil tested and provides you with a copy, contact the soil testing company and go over the report in detail with them. Request that they give you a letter of recertification on the soils, or order a new soil test.

26 DEED RESTRICTIONS

Be aware that a deed restriction placed on property 100 years ago could still be applicable today, and can seriously prohibit the development of your site. A title search by an attorney or a reputable title company should turn up any deed restrictions.

27 EASEMENTS

There could be an easement across your property where a utility line is buried which says you cannot build a building on the easement or within a certain distance. In most cases, of course, this easement is going to be right where you want to put your building or loading dock. Your title search and survey should pick up any easements.

28 FUTURE ASSESSMENTS

If there is a road planned, a new traffic signal required, or new utilities needed, you could get caught having to pay your proportionate share of it as a tax assessment. (Make sure your feasibility study works for you and doesn't wind up looking like the national debt.)

29 ACCELERATION AND DECELERATION LANES

Check with the department of transportation to see if you will need to add acceleration or deceleration lanes to the existing highway as a result of your project. If it is necessary, you need to add it to your feasibility study. It is also usually easier, and less costly, to add during development, than to go ahead with the project only to find you cannot open for business because the department of transportation won't approve it until you make your road adjustments.

30 OTHER USES

If you are an out-parcel to an office building or shopping center, find out what else is going into the center as well as on the other out-parcels. Determine if they will compliment or hurt your operation. You will also want to know this information if you are *in* the shopping center or office building. If you feel there is only room for one of your type of business, ask the developer or leasing agent to give you a written agreement that restricts a direct competitor from coming into the project.

31 VISIBILITY

You have just *got* to be seen. Your client or customer can't do business with you if he cannot find you. Every time you are seen, it registers you and your name with the client or customer.

32 WHAT'S NEXT DOOR

Check out your neighbors. Is the property next door a positive or negative? For example, a residential area might not be too popular next door to a county jail facility. Check with the planning and zoning department to see if there are any projects in the planning stages in the area of your site. Determine whether the neighborhood is declining or improving.

33 NEW GROWTH AREA

New growth areas are a lot of fun but because of the rapid growth taking place, they can get saturated prematurely. Don't be "too much, too soon." Don't go charging in with the cavalry until the settlers get there.

34 REMODELING

If your plan is to remodel, do it well. Don't try to cut corners. It is critical that you establish a whole new image and identity for yourself. And while you are at it, make sure you determine why the previous occupant did not make it, or is no longer in business. Review Chapter 8 on Remodeling.

35 PREMATURE OPENING

Pioneers are out of style. Coordinate your opening with any project or projects that will establish your area as a Destination Point. For example, if your business is a shopping center, planned because of a large residential development under construction, wait until enough of the units are completed and occupied to provide the customers you need. If your business is an industrial plant, planned because of a new road artery under construction, or a new waterway opening up, wait until the roads are completed to give you the access you need. If your business is a restaurant going on an outparcel to a shopping center, coordinate your opening with that of the shopping center.

36 BARRIERS

Natural barriers (rivers, mountains, or a state forest) or man-made barriers (bridges, interstates and railroads) can become like a brick wall where people are concerned. Any one of these barriers can bring an active area to an abrupt stop, or possibly change the entire atmosphere of a town. Plan your project for the side of the barrier that best fits your needs.

37 CURB CUTS

The zoning and planning department and the department of transportation determine how you will access your property. A curb cut is the entrance to your property or site from a city, county or state road. You cannot put them just anywhere, so add this to your list of items to check.

38 NEW ROAD

If your project is going into an area where a new road has just been completed, make sure it is up to standards and has been dedicated to and *accepted by* the appropriate city or county. If it is not up to par, you could be assessed for the corrections or completion.

39 LEASED PROPERTY

If you are buying improved property (property with a building or buildings on it) which is occupied by a tenant, I would suggest very strongly that you do not have your closing until the tenants have vacated the property. If you are not planning to use this property for a long time, of course, you might want them to stay to produce an income for you. If you are planning to use your property right away, however, it can sometimes be a problem getting someone else's tenant to leave. This can also be true of a vacant piece of property that might have ''squatters'' on it. Request that the seller have the property free of all tenants or occupants before closing, and save yourself some real headaches as well as expenses.

40 TITLE COMMITMENT

Title Insurance companies research the seller's title and write an insurance policy that provides coverage for any problem that might come up at some future date that could affect the title. Ask the seller to provide you with a title commitment (a commitment from the title insurance company which says the title is good and they *will* insure it) as soon as possible (within 15 days of contract). Otherwise, you could go to the expense of surveying, engineering, and testing the soil only to find out a week before closing that the seller has a problem on the title and cannot sell you the property.

41 AGREEMENTS

Words are cheap. Only *written* agreements are recognized, so get it *all* in writing.

42 GET A SURVEY

If there has been a recent survey (within 3 months) by a *certified* surveyor, this should be adequate. Make sure, however, that the legal description on your contract matches the one on the survey so that a couple of feet of depth, amounting to 3 or 4 acres, has not been left out.

Also, if there have been any additions to the building, a fence installed or landscaping changed, a new survey would pick up any encroachments. An encroachment is when you have put part of your building on your neighbor's property, or vice versa.

43 FEASIBILITY STUDY

Watch your expenses and estimate your projected sales or income realistically. Underestimating your costs, being misinformed, or too optimistic about your future sales or income for your project, can result in financial ruin.

44 OLD GARBAGE OR CHEMICAL WASTE DUMPS

Have the company that does your soil test do a thorough test. Sometimes, old garbage dumps and/or chemical waste dumps are plowed under and top soil put on top of them. As the garbage decays over time, it causes settlement of the ground. If you have a building on this property, this settlement could cause your building to cave in or collapse. Chemical waste can produce toxic fumes or contaminate the drinking water, among other disasters.

45 ACCESS ROAD OR FRONTAGE ROAD

Some areas are using a frontage road that runs along the major highway with limited entrances or access. This is done to control the flow of traffic entering and leaving a busy road. A frontage road can be okay *but*, if you have come along after a new ruling that all new development must use the frontage road and your competition is sitting there with direct access to the road, you could be overlooked. Positioning to the ingress and egress of the main road is important when a frontage road must be used.

46 IMPROPER CONTRACT

Be sure to include all the clauses you need in your contract. If you need an easement and the seller is to provide it, add that clause to your contract. Have all clauses, agreements and contingencies in your contract.

47 DEVELOPMENT STANDARDS

Be sure to check the requirements for developers and developments with *all* agencies. The water district, environmental agency, tree and wildlife preservation board, as well as the local planning and zoning board requirements must be met. Other agencies could also be involved, so make sure an agency you are not aware of doesn't rear its ugly head and become a dragon.

48 GRANDFATHERED-IN ZONING

If you are purchasing a building that is on property with grandfathered-in zoning, discuss your options under this zoning at length with the zoning department. Most of the time, if you change the use, or the building should burn down (or even if a large percent of it should burn), it is possible that you won't be able to rebuild because you would have to rebuild according to the *new* zoning rules.

49 ADDRESS THE INSURANCE

If your contract is full of contingencies, there is a long period of time until closing, and you will be doing work on the property before closing, specify how insurance will be carried and by whom. Also, be sure to insure the property at closing. Get advice from your insurance agent.

50 DON'T GET EMOTIONAL

The worst thing you can do is fall in love with a piece of real estate and want it no matter what. When this happens, all your logic and business sense goes out of the window. Keep a good business head on your shoulders and look at the facts and feasibility of any project.

MORE FLAGS TO WATCH OUT FOR

Cemeteries—One old family grave in the middle of your property can absolutely ruin your site plan.

Platting—If you are buying a parcel of a larger piece of property that has been platted into several parcels, make sure the plat has *final* approval. A preliminary approval could be affected by changes in zoning or building codes.

Financing—If you are going to need to secure a loan be careful to structure your contract subject to your financing. If you do not add this contingency to your contract, and are unable to obtain financing, you will lose your earnest money or escrow deposit.

Permits—If possible, try to have all of your permits in hand before closing on your site.

Historic Structures—Buying a piece of property that qualifies as a historic structure which cannot be torn down, removed, or changed, can turn out to be a real historical event in your life. See the glossary for the definition of a historic structure.

Archeologic Finds—A good soil test might be able to prevent you from buying a piece of property only to later discover with just one bulldozer ride, that you have uncovered the "archeological find of the century." This puts your whole project on hold until the "find" is worked, making your holding costs sky high.

Wildlife Habitat—The habitat of an endangered species cannot be developed. Know your property *before* you close.

Mineral Rights—If someone owns the mineral rights on your site, it can make it impossible to obtain a loan using the property as collateral because the mortgage is second to the mineral rights.

Improper Signature—Make sure the person who has signed your purchase contract is the legal owner. If the owner is a corporation or partnership, have your attorney verify that the person who signed the purchase agreement has the authority to do so. You could spend money on engineering, soil tests, etc., and have the proper owner sell it right out from under your nose.

Trees Next Door—Trees on adjoining property can block your visibility or the visibility of your sign. Deal with this before you buy the property and get any agreements with adjoining property owners in writing.

Sinkholes—A soil analysis map might be able to alert you to areas prone to sinkholes.

MORE FLAGS

These are just some of the problems I have run into. Any one of these flags can cause problems for you, your project, or your client. You will want to add to this list any other problems you might have seen or heard of to safeguard yourself against as many unknowns as possible. Use the flag sheet provided in this chapter to write them down. Problems that you know about and can address are one thing—those unknowns can be treacherous. The kind of surprises I like are phone calls from a friend or roses from my sweetheart. A call from your surveyor, *after closing* saying you have bought 42 acres instead of 50 can ruin a perfectly nice day.

14

Professional Services

Yaba Daba Do! You, your real estate broker, and your attorney have negotiated a good contract with the seller and you are off and running. Remember: honesty, fairness, and integrity? They still work and are even back in style.

Now it's time to get to work on the site itself to see if it is all going to work for you. There is only one way to do this and that is to hire professionals to get the job done. Check references with the people you intend to hire, and don't pinch pennies. Hire the best.

In this chapter, I've outlined some of the certified professionals you will want to consider hiring for your project, and what they can do for you.

ACCOUNTANT

A Certified Professional Accountant (CPA) is a person educated and certified by the state to manage business accounts and examine your assets and liabilities to verify your financial worth. A CPA can prepare a feasibility study and advise you on income tax consequences pertaining to you and your project.

The CPA can also prepare financial statements for financing purposes, as well as set up a proper accounting or bookkeeping system for your project.

A certified public accountant can go over your net worth and prepare a feasibility study for you.

In today's business world, feasibility studies, financial statements, profit and loss statements, etc. are only considered to be true and accurate when prepared and certified by a CPA.

APPRAISER

An appraiser is someone trained in appraising *or* setting a value on property. By setting a value on your property the appraiser prepares a written appraisal that usually includes the sales of comparable properties in the area with similar zoning, road frontage, and other variables. Differences in property used in the comparable sales analysis is credited or debited to the value of your property. If the next door property sold for a lower price but did not have as much frontage on a paved road as yours, your property would be credited with better value. If, however, the other property instead of yours had the road frontage and sold for a higher price, then your property would be adjusted lower because your property did not consist of road frontage.

Comparing property sizes and characteristics are all part of the appraisal process. You must use a certified appraiser who has a designation. The M.A.I. certification is awarded by the American Institute of Appraisal and is usually acceptable for any purpose. A bank or lending institution for example, will require an appraisal by a certified appraiser (sometimes two).

ARCHITECT

An architect is a person whose profession is designing and drawing up plans for buildings, and usually supervises the construction.

You want to find an architect who is a member of the American Institute of Architects. This membership is earned only through proper educational and licensing requirements.

An architect can also specialize in city planning, site planning, and landscape design. Of course, they could also specialize in areas that are not relative to what we are discussing.

Permits for building and developing your site are based on plans and site layouts prepared by a certified architect.

ATTORNEY

An attorney is trained in the law and empowered to advise or represent others in legal matters.

Some attorneys specialize in one particular area. An attorney who specializes in real estate and purchasing and selling property will be the most helpful to you. He will know the clauses critical to your success.

An attorney can structure your contract to fit your needs. He can also help you with lease papers, municipal legalities, as well as how to structure your corporation or partnership.

As I mentioned earlier, don't try to remove your own appendix *or* write your own purchase contract.

BUILDING CONTRACTOR

A building contractor is trained and licensed in the construction of buildings. It could be beneficial to use a builder who specializes in the type of project you are doing. Some builders specialize in shopping centers, some in warehouses, etc.

A builder needs to be licensed by the state. A license and certification indicates proper education and knowledge of the building codes and requirements.

CONTRACTOR

A contractor is trained and licensed in a particular field of expertise. You need a specialist contractor for all phases of your development including roads, buildings, utilities, site work, etc.

Some contracting firms are set up with specialists in each field so that you can coordinate your entire project through one office. This works very well and is a real time saver.

FINANCIAL PLANNER

A financial planner is a certified professional who helps you plan your income tax strategies, your estate, and your retirement. Any project you do should be a part of your overall financial plan.

A financial planner might also be able to help you find financing or joint venture partners for your project.

INSURANCE AGENT

An insurance agent is trained and licensed to represent insurance companies in counseling, advising, and selling insurance coverage. You need insurance right from the time you take legal responsibility for work orders and ownership of the property or improvements. Keep your insurance agent informed of your activities at all times, there could be coverage you need that you are not aware of.

LEASING AND MARKETING AGENT

If you plan to construct space to rent to clients or tenants, it could be very helpful to you to hire an experienced, knowledgeable agent who specializes in handling marketing and leasing for you. This person would be responsible for advertising, signage, and negotiating your leases for you. It is best to have your attorney draft your lease for you to make sure you have the protection and clauses you need.

Don't cut yourself short on a good marketing agent. This is a person or company who promotes and advertises your product. A good marketing agent is a valuable asset.

LENDER

The company or person who lends you the money to complete your project plays a very important role. You can work with a lender through your mortgage broker, or you can work directly with the bank or investment company. Discuss your plans with the lender so you will not only know what he expects of you, but also be able to determine if your lender can handle the project in a way that will work for you.

Interest rates, origination fees, points on the loan, length of mortgage, personal guarantees, and construction draws are just some of the topics you want to discuss with a lender.

A lender wants information from you such as a financial statement, purchase contract, appraisal, projections, and a performa of your plans. Be sure you have a package to take to the lender that provides a clear, complete picture of your project.

ROAD PLANNING AND PAVING CONTRACTOR

Road planning and paving contractors are trained and certified professionals in the planning and construction of road systems. They can also design and pave a shopping center or office complex parking lot.

Sometimes these contractors and engineers set up a company together and can coordinate your whole project for you.

SOIL TESTING COMPANY

It is imperative that you get a soil test. A sample soil test report is shown in Fig. 14-2. This should be done by a local company, familiar with the peculiarities of the soils in the area.

The engineering department should be able to give you a good idea of the types of soils in the area, and what you need to

GENERAL SUBSURFACE CONDITIONS

Soil stratification, is based on examination of recovered soil samples and interpretation of field data. The stratification lines represent the approximate boundary between the soil types and the actual transition may be gradual.

The groundwater level was encountered, at the time of our exploration of depths that varied from standing water at the ground surface to a depth of 2.5 feet below the ground surface. The deeper water levels were encountered in the eastern portion of the site and the high (shallow) water table conditions were encountered in the western portion of the site. Fluctuations in all groundwater levels should be anticipated throughout the year primarily due to seasonal variation in rainfall.

In general, the following two distinct areas were encountered during our field exploration program:

Eastern Portion of the Site

The soils in the eastern portion of the site are characterized by a surface deposit of peat underlain by granular soils. The general location of the peat is shown as a shaded area on Figure 2. As indicated in SPT boring TH-1, the soils below the peat consists of slow draining loose to medium dense clayey fine sands to a depth of 12 feet. The clayey fine sand is underlain by loose to medium dense granular soils consisting of fine sand and silty to clayey fine sand. Once the muck has been removed and replaced with clean compacted sand, the site can be used for constructions.

Western Portion of the Site

No unsuitable much was encountered in the western portion of the site. However, access to this area was limited due to the thick dense vegetation and therefore, organic soils may be present. The soils we could sample were wet with a ground water table close to or at the surface. Four shallow hand-augers were conducted in this portion of the site to a maximum depth of 4 feet. Silty to clayey sand was encountered within this 4-foot depth.

Comments

Based on the data obtained from our field exploration, the compressible organic soils "muck" must be completely removed, or "demucked", from the site and replaced with compacted sands prior to construction.

The "demucking" operation should be performed "in-the-dry" so as to properly inspect for the complete removal of the organic material; demucking under wet conditions can often lead to over-excavation of f the "good" material underlying the organic "bad"

Fig. 14-2. *A sample soil report.*

material. Therefore, dewatering as a means to control the water is recommended during the "demucking" operation. Once the organic materials have been removed, the excavation should be backfilled with compacted material.

Soil layer (2), without roots, can be reused for compacted backfill. Soil layer (3) can be used if the moisture content is reduced (dried) below the optimum value.

The entire site should be grubbed of the existing vegetation and roots. The western portion of the site will probably require the addition of fill to improve the drainage. We recommend that clean, free draining, fine sand is used. All fill material should be compacted.

Closure

The information presented herein is based upon our preliminary subsurface soil exploration at the subject rate. The intent of the information presented herein is for a general qualitative assessment of the soil conditions encountered in the accessible portions of the site. A large portion of the site was not accessible due to very dense vegetation and thorned bushes.

We recommend that we conduct a follow-up subsurface soil exploration and analysis once the actual building configuration(s), parking/drive areas all fill requirements have been determined and access to our drilling equipment is provided. The follow-up subsurface soil exploration and analysis should be performed to better define the subsurface soil conditions below the depths (excluding TH-1) explored for this preliminary report and, therefore, to present site-specific site foundation support recommendations.

watch out for and why. Veins of rock deposits, muck, old garbage dumps, toxic waste dumps, old cemeteries, and natural drain areas should be familiar to the local engineering department.

Discuss the soil test with the soil company. If there are any particular tests you want, be sure to go over them. Ask for ideas from the company and, of course, ask lots of questions. Get a clear picture of the why's and wherefores of everything you are doing.

SURVEYOR

A surveyor is trained and certified in the field of surveying. This person knows how to examine your property and determine (or verify) the location, the amount of acreage or square feet, the boundaries, easements or encroachments, and the elevation.

A surveyor can determine your easements, boundaries, encroachments, and the elevation of your site.

The surveyor can also identify and locate any structures on your property as well as any hills, mountains, lakes or depressions, and trees.

The surveyor also provides this information for you in written detail. The surveyor should be certified and licensed by the state, or you could be wasting your money. The city and county departments you will be working with, as well as the lenders, require surveys be certified by a licensed surveyor.

UTILITY ENGINEER

Utilities are the electric, gas, water, and sewer services you need for your project. A utility engineer is a person who is trained to design and construct the utilities for your site or project, utilizing the features you have in setting up your services.

These are some of the professionals you will need to get you started. As time goes on, you might want to add to the list. Be as thorough as possible; ask lots of questions, and above all else, hire qualified, certified professionals.

The length of time it takes to complete a project varies a great deal from project to project. Finding a good site is the foundation of your project, so a quick decision or flipping a coin is out of the question.

After your site is chosen, the purchase contract must be structured and negotiated. After your contract is signed by both parties, you call in the professionals. Zoning, permitting, site work, and construction could take from 60 days to 3 years or more. The professionals you hire can help you determine the time it will take so you can make preliminary schedules, etc.

Try not to box yourself into a time frame that is not adequate. For instance, if you promise your seller you will close in 90 days when you get your permit and financing—make sure that is enough time, otherwise it could void your contract, or cost you money for an extension. The dates and times on your purchase agreement are important and should be met.

Sometimes the seller allows you all the time you need, but might charge you for it. For instance, you and the seller sign a contract on a site subject to site plan approval and it appears it will take six to eight months to get it. The seller agrees with the time period, but asks you to pay his holding costs of 1 percent a month. If you close earlier to save this cost, you could find yourself owning a site you cannot use because the site plan did not work.

You will want to move your project along as much as you can, but don't rush it along at a speed that allows costly mistakes.

15

Closing the Deal

By now, you know your ABC's: you have identified your town, **A**rea or section, found your **B**est site, and analyzed your **C**ompetition.

You are doing great. You've left something out, however. Where are you going with this project without a decision?

THE DECISION

If your project is for your boss or client, you need to obtain a decision from them. You can make a choice by listing the sites you would want as 1, 2, and 3. One being the best or first choice. But don't get too far off the point. The point being what *your* site needs to be: Better, Better, Better.

Once you have made the decision to go with a particular site, you have the name, address, and phone number of the owner and have been given the asking price. You are ready to make the owner (now he becomes a seller) an offer.

THE CONTRACT

Have an attorney, preferably a real estate attorney, write a contract for you. I don't mean to drum up business for the legal profession, however, you would not try to remove your own appendix, why would you write your own contract?

An attorney can advise you of any special clauses, as well as standard clauses, that need to be included with your offer, depending on the type of property you are buying, what you plan to do with it, and when you plan to do it.

It is important to you that your purchase contract reflect your development plans. Don't box yourself into deadlines and commitments you cannot meet. You can be sued for nonperformance, lose your earnest money deposit, lose the money you have spent so far on site work, permitting, etc., or all of the above.

CLAUSES

Spend time with your attorney discussing your plans for your project. Explain to your attorney why you are buying this property and what you plan to do with it so he has a clear enough picture to draft your contract to protect your interests. A good real estate attorney should have a computer full of clauses to be included in your contract for your protection.

The following are some contingencies you may want to include in your contract, depending on your plans for development. If these contingencies are not met, it voids your contract, you will not be expected to buy the property and your earnest money deposit should be returned to you.

Time

You will need time to get everything done. If possible, provide for time extensions in case of a snag that holds you up. Rezoning or environmental approvals can take a very extended period of time.

Permits

If your project needs permits, have your contract contingent upon obtaining these permits. Address *all* the permits you need. As sure as you leave one off, that will be the one you will have a problem with.

A good real estate attorney should have a computer full of clauses to be included in your contract for your protection.

Utilities

You want to include a clause in your contract that allows you to verify where the utility lines are, if they are adequate, and if there is enough utility capacity for you to use. You are planning to develop this property, if you cannot get water or sewer because there is no capacity left in the area, you might want to find other property that does have adequate capacity.

Soils

Your contract should be subject to favorable soil tests, and provide for a remedy if the soil test is not favorable. The remedy should spell out what choices could be taken if the soil tests are not favorable. For example, if the cost to correct the soil problem can be determined, the buyer and seller could elect to split this cost up to a percentage of the purchase price or a set dollar amount.

Permission

The contract should provide permission for you and/or your contractors to enter the property for engineering, testing and so forth, as well as who is responsible for any liability during this period of time. The seller will probably want it made clear that the costs of these tests are not his.

Signs

Your contract should allow you to erect signs as needed as long as they comply with the local sign ordinance.

Financing

If you are going to finance your project, your contract should have a contingency clause for financing. The lender will want to see your engineering, soil tests, architectural plans, etc., and a copy of your contract. Therefore, you will not be able to obtain financing before you negotiate your contract.

Design

If you are purchasing the property for a specific purpose or special design such as a 300,000 square foot shopping center, a pizza parlor, or an industrial warehouse of a specific size, make sure your contract is contingent on this special design fitting the property. The zoning, building, parking, and utility requirements for this specific use must be met. You might not be able to get approvals for your pizza parlor because you are short five parking spaces and the site is not large enough.

Personal Approvals

If you need approvals from your partner, a committee—perhaps your Franchiser—add a contingency to include it.

Title

Always make your contract contingent upon the seller providing a clear and marketable title. If there should be a problem, address the remedy. For example, seller will have 15 days to clear a defect.

Survey

If a current boundary and topographical survey has not been completed (within 30 days of contract), the contract should be contingent upon these surveys. A remedy should be stated if a problem should arise, such as removal of enroachments within a set number of days.

Possession

If the property has tenants, or is occupied in any way, how and when the property will be vacated should be outlined.

Cooperating Seller

The contract should provide for permission and cooperation of the seller if needed, to obtain zoning changes, municipality changes (annexation into the city limits, for example), easements, dedications, vacations, petitions, applications, road changes for ingress and/or egress, etc.

There are, of course, a lot of other standard contract clauses that must be in the contract. These are some of the more commonly used contingency clauses.

PRESENTATION AND NEGOTIATIONS

When you and your attorney have completed the contract, it should be presented to the seller for acceptance. If you have been working with a real estate broker, he can present the contract to the seller and negotiate the terms and conditions for you. The broker who has the CCIM designation is trained as a negotiator and would be a valuable help to you in finalizing the contract you will need to work with on your project.

If you have not worked with a real estate broker, your attorney can present the contract for you. I find that a third party negotiator can be a tremendous asset.

If your offer or terms are not acceptable to the seller, your negotiator should be able to determine what will be acceptable. He can then get back with you and your attorney to determine acceptability of this counteroffer. This process can sometimes rock back and forth until a suitable agreement or "meeting of the minds" is accomplished. And, of course, sometimes an agreement is not possible because of unavoidable need differences of the buyer and seller.

With permits in hand, financial approval and commitment, title work cleared, and any and everything completed and checked for your project, your next step is to close on the property and start construction.

A schedule is set up and your leasing, marketing, promotions, and grand opening are in place. If necessary, you begin to plan your staffing and training. Your project is on its way to completion.

Good luck—see you at the Opening.

Appendix A

Sample Demographic Report

Prepared by National Decision Systems

DESCRIPTION	3.0 MILE RADIUS	5.0 MILE RADIUS	10.0 MILE RADIUS
POPULATION			
1993 PROJECTION	41,949	117,398	470,638
1988 ESTIMATE	38,186	103,121	410,767
1980 CENSUS	32,395	81,011	318,765
1970 CENSUS	19,880	55,167	243,476
GROWTH 70-80	62.95%	46.85%	30.92%
HOUSEHOLDS			
1993 PROJECTION	16,229	45,346	182,185
1988 ESTIMATE	14,669	39,552	157,316
1980 CENSUS	12,057	30,120	117,185
1970 CENSUS	5,790	16,619	78,175
GROWTH 70-80	108.26%	81.23%	49.90%
POPULATION BY RACE & SPANISH ORIGIN	32,395	81,011	318,765
WHITE	90.61%	85.71%	80.81%
BLACK	5.37%	11.37%	16.82%
AMERICAN INDIAN	0.59%	0.48%	0.32%
ASIAN & PACIFIC ISLANDER	1.10%	1.04%	0.86%
OTHER RACES	2.33%	1.39%	1.19%
SPANISH ORIGIN - NEW CATEGORY	8.30%	5.48%	4.24%
OCCUPIED UNITS	12,057	30,120	117,185
OWNER OCCUPIED	51.10%	55.16%	59.70%
RENTER OCCUPIED	48.90%	44.84%	40.30%
1980 PERSONS PER HOUSEHOLD	2.68	2.66	2.61
YEAR ROUND UNITS AT ADDRESS	13,023	32,857	126,206
SINGLE UNITS	66.75%	71.96%	72.58%
2 TO 9 UNITS	19.21%	14.81%	12.68%
10+ UNITS	9.25%	8.18%	8.93%
MOBILE HOME OR TRAILER	4.78%	5.05%	5.81%
SINGLE/MULTIPLE UNIT RATIO	2.35	3.13	3.36
1988 ESTIMATED HOUSEHOLDS BY INCOME	14,669	39,552	157,316
$75,000 OR MORE	5.39%	9.64%	7.94%
$50,000 TO $74,999	14.72%	17.27%	14.81%
$35,000 TO $49,999	21.13%	19.40%	17.20%
$25,000 TO $34,999	18.59%	16.96%	16.71%
$15,000 TO $24,999	20.63%	18.10%	19.06%
$7,500 TO $14,999	10.72%	10.09%	12.78%
UNDER $7,500	8.82%	8.54%	11.14%
1988 ESTIMATED AVERAGE HH INCOME	$33,727	$38,472	$34,846
1988 ESTIMATED MEDIAN HH INCOME	$30,938	$34,569	$30,564
1988 ESTIMATED PER CAPITA INCOME	$13,048	$14,928	$13,594

DESCRIPTION	3.0 MILE RADIUS	5.0 MILE RADIUS	10.0 MILE RADIUS
POPULATION BY SEX	32,388	80,977	318,698
MALE	49.59%	49.20%	48.52%
FEMALE	50.41%	50.80%	51.48%
POPULATION BY AGE	32,388	80,977	318,698
UNDER 5 YEARS	6.45%	6.59%	6.45%
5 TO 9 YEARS	6.87%	7.00%	6.77%
10 TO 14 YEARS	7.63%	7.61%	7.34%
15 TO 19 YEARS	10.60%	9.50%	10.20%
20 TO 24 YEARS	12.87%	11.47%	10.82%
25 TO 29 YEARS	10.71%	10.16%	9.27%
30 TO 34 YEARS	8.00%	8.22%	7.51%
35 TO 44 YEARS	11.70%	11.66%	10.57%
45 TO 54 YEARS	11.31%	11.15%	10.50%
55 TO 59 YEARS	4.98%	5.25%	5.37%
60 TO 64 YEARS	3.46%	3.94%	4.45%
65 TO 74 YEARS	3.68%	4.73%	6.48%
75+ YEARS	1.75%	2.72%	4.27%
MEDIAN AGE	28.13	29.85	31.42
AVERAGE AGE	31.29	32.62	34.20
FEMALE POPULATION BY AGE	16,326	41,138	164,065
UNDER 5 YEARS	6.15%	6.28%	6.14%
5 TO 9 YEARS	6.97%	6.85%	6.49%
10 TO 14 YEARS	7.46%	7.23%	6.93%
15 TO 19 YEARS	10.68%	9.41%	9.33%
20 TO 24 YEARS	12.92%	11.62%	10.51%
25 TO 29 YEARS	10.40%	9.85%	8.96%
30 TO 34 YEARS	7.63%	7.95%	7.35%
35 TO 44 YEARS	11.89%	11.74%	10.61%
45 TO 54 YEARS	11.30%	11.17%	10.64%
55 TO 59 YEARS	4.97%	5.27%	5.60%
60 TO 64 YEARS	3.38%	4.01%	4.67%
65 TO 74 YEARS	4.05%	5.06%	7.23%
75+ YEARS	2.19%	3.56%	5.55%
FEMALE MEDIAN AGE	27.83	29.14	29.86
FEMALE AVERAGE AGE	31.74	33.42	35.68
POPULATION BY HOUSEHOLD TYPE	32,388	80,977	318,698
FAMILY HOUSEHOLDS	84.91%	84.43%	82.00%
NON FAMILY HOUSEHOLDS	14.93%	14.47%	13.95%
GROUP QUARTERS	0.17%	1.10%	4.04%

DESCRIPTION	3.0 MILE RADIUS	5.0 MILE RADIUS	10.0 MILE RADIUS
HISPANIC POPULATION BY RACE	2,689	4,441	13,531
WHITE	75.65%	75.53%	71.79%
BLACK	1.27%	3.23%	5.62%
AMERICAN INDIAN & ASIAN	1.01%	2.46%	1.91%
OTHER RACE	22.07%	18.77%	20.69%
HISPANIC POPULATION BY TYPE	32,395	81,011	318,765
NOT OF HISPANIC ORIGIN	91.70%	94.52%	95.76%
MEXICAN	0.42%	0.46%	0.51%
PUERTO RICAN	3.53%	2.23%	1.51%
CUBAN	2.38%	1.33%	1.07%
OTHER HISPANIC	1.97%	1.46%	1.14%
MARITAL STATUS PERSONS 15+	25,604	63,815	253,181
SINGLE	28.04%	26.02%	26.96%
MARRIED	55.86%	56.83%	53.69%
SEPARATED	2.29%	2.43%	2.80%
WIDOWED	4.07%	5.27%	7.40%
DIVORCED	9.74%	9.45%	9.15%
MARITAL STATUS FEMALES 15+	12,965	32,766	131,964
SINGLE	24.36%	22.74%	22.83%
MARRIED	55.08%	55.27%	51.40%
SEPARATED	2.52%	2.58%	3.13%
WIDOWED	6.93%	8.79%	12.18%
DIVORCED	11.11%	10.62%	10.46%
PERSONS IN UNIT	12,057	30,115	117,172
1 PERSON UNITS	20.94%	21.67%	24.08%
2 PERSON UNITS	33.52%	34.53%	34.31%
3 PERSON UNITS	19.77%	18.39%	17.34%
4 PERSON UNITS	14.63%	14.23%	13.24%
5 PERSON UNITS	6.77%	6.65%	6.30%
6+ PERSON UNITS	4.37%	4.53%	4.74%
PERSONS IN RENTER UNITS	5,882	13,472	47,179
1 PERSON UNITS	31.59%	30.99%	35.04%
2 PERSON UNITS	35.58%	34.19%	31.76%
3 PERSON UNITS	17.65%	16.65%	15.14%
4 PERSON UNITS	9.03%	10.53%	9.70%
5 PERSON UNITS	3.69%	4.50%	4.52%
6+ PERSON UNITS	2.47%	3.14%	3.85%

DESCRIPTION	3.0 MILE RADIUS	5.0 MILE RADIUS	10.0 MILE RADIUS
HOUSEHOLDS BY TYPE	12,057	30,115	117,172
SINGLE MALE	11.11%	10.65%	9.38%
SINGLE FEMALE	9.83%	11.02%	14.70%
MARRIED COUPLE	57.17%	57.86%	55.35%
OTHER FAMILY - MALE HEAD	3.05%	2.86%	2.79%
OTHER FAMILY - FEMALE HEAD	10.23%	9.94%	11.66%
NON FAMILY - MALE HEAD	5.54%	4.98%	3.76%
NON FAMILY - FEMALE HEAD	3.07%	2.70%	2.37%
HOUSEHOLDS WITH CHILDREN 0-18	4,859	11,624	42,808
MARRIED COUPLE FAMILY	75.58%	75.35%	71.93%
OTHER FAMILY - MALE HEAD	4.19%	4.18%	4.03%
OTHER FAMILY - FEMALE HEAD	18.70%	18.89%	22.67%
NON FAMILY	1.53%	1.59%	1.37%
1980 OWNER OCCUPIED PROPERTY VALUES	5,249	13,545	57,161
UNDER $25,000	8.33%	11.57%	14.55%
$25,000 TO $39,999	32.58%	26.51%	30.84%
$40,000 TO $49,999	27.09%	19.46%	19.21%
$50,000 TO $79,999	25.11%	29.06%	25.37%
$80,000 TO $99,999	3.29%	6.50%	4.61%
$100,000 TO $149,999	2.56%	4.86%	3.75%
$150,000 TO $199,999	0.58%	1.23%	0.93%
$200,000+	0.45%	0.81%	0.75%
1980 MEDIAN PROPERTY VALUE	$45,471	$50,333	$45,849
POPULATION BY URBAN VS RURAL	32,395	81,011	318,765
URBAN	99.58%	99.15%	94.83%
RURAL	0.42%	0.85%	5.17%
POPULATION ENROLLED IN SCHOOL	8,567	21,145	80,667
NURSERY SCHOOL	4.06%	4.39%	4.24%
KINDERGARTEN & ELEMENTARY (1-8)	52.13%	50.41%	51.45%
HIGH SCHOOL (9-12)	23.86%	24.77%	25.03%
COLLEGE	19.94%	20.44%	19.28%
POPULATION 25+ BY EDUCATION LEVEL	17,986	46,904	186,237
ELEMENTARY (0-8)	9.87%	10.39%	13.49%
SOME HIGH SCHOOL (9-11)	14.31%	13.97%	15.39%
HIGH SCHOOL GRADUATE (12)	44.11%	39.83%	37.46%
SOME COLLEGE (13-15)	18.47%	19.63%	18.64%
COLLEGE GRADUATE (16+)	13.24%	16.17%	15.01%

DESCRIPTION	3.0 MILE RADIUS	5.0 MILE RADIUS	10.0 MILE RADIUS
POPULATION 16+ BY OCCUPATION	17,835	41,200	147,960
EXECUTIVE AND MANAGERIAL	10.50%	11.84%	11.18%
PROFESSIONAL SPECIALTY	9.58%	11.49%	11.24%
TECHNICAL SUPPORT	2.88%	3.09%	2.98%
SALES	11.35%	11.20%	11.69%
ADMINISTRATIVE SUPPORT	19.66%	18.54%	17.92%
SERVICE: PRIVATE HOUSEHOLDS	0.38%	0.46%	0.78%
SERVICE: PROTECTIVE	1.68%	1.87%	1.79%
SERVICE: OTHER	15.83%	15.40%	15.02%
FARMING FORESTRY & FISHING	1.51%	1.51%	2.12%
PRECISION PRODUCTION & CRAFT	11.62%	10.76%	11.68%
MACHINE OPERATOR	5.53%	5.16%	4.90%
TRANS. AND MATERIAL MOVING	5.29%	4.53%	4.31%
LABORERS	4.18%	4.15%	4.37%
FEMALES 16+ WITH CHILDREN 0-18	4,489	10,724	39,759
WORKING WITH CHILD UNDER 6	22.41%	22.89%	23.86%
NOT WORKING WITH CHILD UNDER 6	15.07%	17.75%	18.34%
WORKING WITH CHILD 6-18	44.20%	40.75%	39.67%
NOT WORKING WITH CHILD 6-18	18.33%	18.60%	18.13%
HOUSEHOLDS BY NUMBER OF VEHICLES	12,034	30,017	117,112
NO VEHICLES	4.45%	5.24%	9.71%
1 VEHICLE	40.67%	41.90%	41.67%
2 VEHICLES	36.62%	36.06%	33.64%
3+ VEHICLES	18.27%	16.81%	14.98%
ESTIMATED TOTAL VEHICLES	20,742	50,366	183,741
POPULATION BY TRAVEL TIME TO WORK	17,812	42,014	153,918
UNDER 5 MINUTES	3.15%	2.64%	3.37%
5 TO 9 MINUTES	15.24%	12.21%	12.93%
10 TO 14 MINUTES	17.67%	16.80%	16.46%
15 TO 19 MINUTES	18.85%	20.54%	19.87%
20 TO 29 MINUTES	23.05%	24.85%	24.32%
30 TO 44 MINUTES	16.47%	17.41%	16.67%
45 TO 59 MINUTES	2.70%	2.54%	3.03%
60+ MINUTES	2.88%	3.01%	3.35%
AVERAGE TRAVEL TIME IN MINUTES	19.00	19.68	19.72
POPULATION BY TRANSPORTATION TO WORK	17,579	41,414	154,928
DRIVE ALONE	74.10%	72.04%	66.81%
CAR POOL	18.68%	20.08%	19.61%
PUBLIC TRANSPORTATION	1.52%	2.12%	2.57%
WALKED ONLY	2.37%	2.19%	6.63%
OTHER MEANS	2.71%	2.81%	3.26%
WORKED AT HOME	0.62%	0.76%	1.13%

DESCRIPTION	3.0 MILE RADIUS	5.0 MILE RADIUS	10.0 MILE RADIUS
HOUSING UNITS BY YEAR BUILT	13,008	32,780	126,187
BUILT 1979 TO MARCH 1980	2.94%	5.74%	5.30%
BUILT 1975 TO 1978	10.94%	11.28%	9.50%
BUILT 1970 TO 1974	39.52%	31.00%	23.46%
BUILT 1960 TO 1969	29.66%	29.13%	25.62%
BUILT 1950 TO 1959	14.21%	16.06%	22.04%
BUILT 1940 TO 1949	1.45%	3.84%	7.40%
BUILT 1939 OR EARLIER	1.27%	2.96%	6.68%
1980 HOUSEHOLDS BY 1979 INCOME	12,057	30,120	117,178
$75,000+	0.69%	1.24%	1.09%
$50,000 TO $74,999	1.91%	2.75%	2.32%
$35,000 TO $49,999	6.67%	7.79%	6.39%
$25,000 TO $34,999	14.22%	14.05%	12.97%
$15,000 TO $24,999	31.73%	30.16%	27.59%
$7,500 TO $14,999	28.58%	26.87%	27.45%
UNDER $7,500	16.21%	17.16%	22.19%
1979 AVERAGE HOUSEHOLD INCOME	$18,610	$19,726	$18,324
1979 MEDIAN HOUSEHOLD INCOME	$16,804	$17,322	$15,662
1980 FAMILIES BY 1979 INCOME	8,533	21,334	82,158
$75,000+	0.76%	1.37%	1.72%
$50,000 TO $74,999	2.12%	3.25%	2.92%
$35,000 TO $49,999	8.40%	9.88%	7.96%
$25,000 TO $34,999	16.41%	16.63%	16.01%
$15,000 TO $24,999	34.04%	32.08%	30.86%
$7,500 TO $14,999	25.92%	24.53%	25.49%
UNDER $7,500	12.35%	12.25%	15.04%
1979 AVERAGE FAMILY INCOME	$20,218	$21,739	$21,175
1979 MEDIAN FAMILY INCOME	$18,487	$19,419	$18,211

DESCRIPTION	3.0 MILE RADIUS	5.0 MILE RADIUS	10.0 MILE RADIUS
1988 POPULATION BY SEX	38,186	103,121	410,767
MALE	49.37%	48.83%	48.54%
FEMALE	50.63%	51.17%	51.46%
1988 POPULATION BY AGE	38,186	103,121	410,767
UNDER 5 YEARS	6.46%	6.69%	6.80%
5 TO 9 YEARS	6.44%	6.59%	6.62%
10 TO 14 YEARS	6.02%	6.18%	6.10%
15 TO 19 YEARS	7.66%	7.22%	7.78%
20 TO 24 YEARS	10.55%	9.22%	9.00%
25 TO 29 YEARS	10.49%	9.90%	9.35%
30 TO 34 YEARS	9.31%	9.18%	8.82%
35 TO 44 YEARS	15.32%	15.13%	14.36%
45 TO 54 YEARS	9.85%	9.91%	9.34%
55 TO 59 YEARS	4.98%	5.00%	4.89%
60 TO 64 YEARS	4.46%	4.72%	4.82%
65 TO 74 YEARS	5.94%	6.89%	7.41%
75+ YEARS	2.52%	3.36%	4.71%
1988 MEDIAN AGE	31.54	32.73	33.24
1988 AVERAGE AGE	33.99	35.02	35.72
1988 FEMALE POPULATION BY AGE	19,334	52,767	211,378
UNDER 5 YEARS	6.17%	6.39%	6.43%
5 TO 9 YEARS	6.13%	6.29%	6.26%
10 TO 14 YEARS	5.83%	5.90%	5.78%
15 TO 19 YEARS	7.92%	7.17%	7.21%
20 TO 24 YEARS	10.65%	9.39%	8.77%
25 TO 29 YEARS	9.96%	9.51%	8.95%
30 TO 34 YEARS	8.91%	8.88%	8.56%
35 TO 44 YEARS	15.32%	15.00%	14.23%
45 TO 54 YEARS	10.12%	10.09%	9.49%
55 TO 59 YEARS	5.05%	5.07%	5.02%
60 TO 64 YEARS	4.56%	4.86%	5.11%
65 TO 74 YEARS	6.20%	7.30%	8.16%
75+ YEARS	3.18%	4.14%	6.03%
1988 FEMALE MEDIAN AGE	32.15	33.57	34.97
1988 FEMALE AVERAGE AGE	34.70	35.96	37.42

ORANGE BLOSSOM TRAIL AND SAND LAKE ROAD — Three mile Radius
ORLANDO, FLORIDA BY CENSUS TRACT

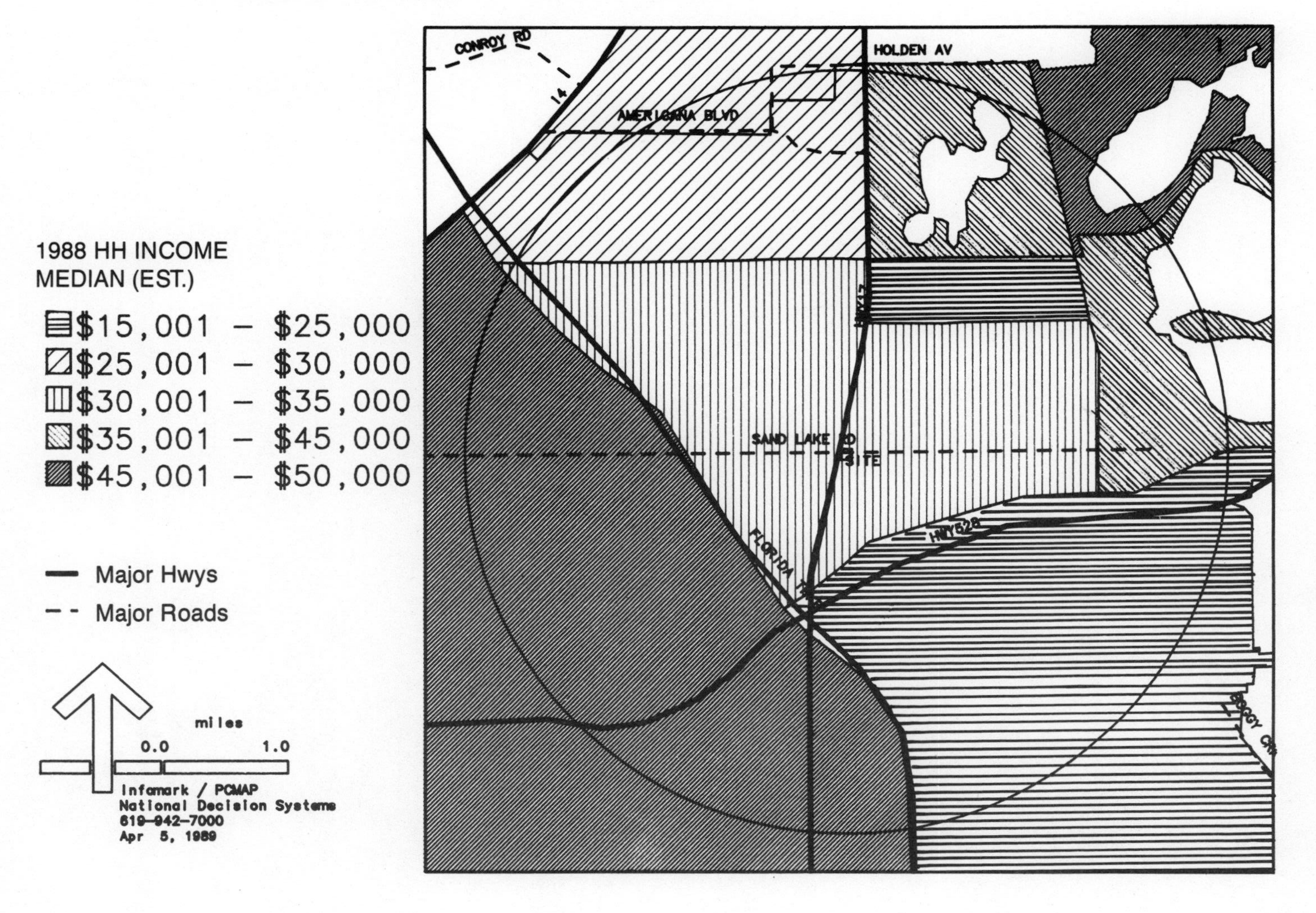
CONROY RD
HOLDEN AV
AMERICANA BLVD
SAND LAKE RD
SITE
FLORIDA TPK
HWY528
BOGGY CR
1988 HH INCOME
MEDIAN (EST.)
$15,001 — $25,000
$25,001 — $30,000
$30,001 — $35,000
$35,001 — $45,000
$45,001 — $50,000
Major Hwys
Major Roads
miles
0.0
1.0
Infomark / PCMAP
National Decision Systems
619-942-7000
Apr 5, 1989

Appendix B

Sample Zoning Sheets

COMMERCIAL GENERAL DISTRICT

(C G District)

A. INTENT:

This district is intended for general commercial activity. Businesses in this category require larger land area and location convenient to automotive traffic. Pedestrian traffic will be found in this district. The district is not suitable for heavily automotive-oriented uses. It is not the intent of this district that it shall be used to encourage extension of strip commercial areas. Multi-family dwellings and transient accommodations are permitted by special exception. It is generally intended to utilize this district to implement the Sarasota County Comprehensive Plan, *Apoxsee*, within those areas of Sarasota County shown as "Designated Urban Areas," and more specifically, the "Town, Community, and Village Activity Centers" as shown on the "Future Land Use" Plan Map. Whenever transient accommodations are involved, maximum density shall be guided by the Urban Area Residential Checklist and the Urban Area Residential Intensity Matrix contained within the "Future Land Use" Plan Chapter's Guiding Principles of *Apoxsee*. (ORD. 83-08, 2/15/83)

B. PERMITTED PRINCIPAL USES AND STRUCTURES:

1. Retail outlets for sale of food, wearing apparel, toys, sundries and notions, books and stationery, leather goods and luggage, jewelry (including watch repair but not pawnshop), art, cameras or photographic supplies (including camera repair), sporting goods, hobby shops, musical instruments, television and radio (including repair incidental to sales), hardware*, florist or gift shop, delicatessen, bake shop (but not wholesale bakery), drugs, and similar products. (*ORD. 76-71, 8/14/76)
2. Service establishments such as barber or beauty shop, shoe repair shop, restaurant (but not drive-in fast food* restaurant), photographic studio, dance or music studio, self service laundry, tailor, draper or dressmaker, laundry or dry cleaning pick-up station, and similar activities. (*ORD. 78-15, 5/29/79)
3. Small loan agencies, travel agencies, employment offices, newspaper office (but not printing or circulation), and similar establishments.
4. Professional and business offices, and medical or dental clinic.
5. Private clubs and lodges, and libraries.
6. House of worship.
7. Railroad rights-of-way.

8. Retail outlets for sale of home furnishings and appliances (including repair incidental to sales), office equipment or furniture, antiques, pet shop and grooming (but not animal kennel), automotive parts (including installation)* and similar uses. (*ORD. 76-72, 8/24/76)
9. Service establishments such as radio or television station (but not transmitter-tower), funeral home, interior decorator, upholstery shop (but not furniture refinishing*), marina, radio and television repair shop, health spa, letter shops and printing establishments not involving linotype or large scale type-setting, and similar uses. (*ORD. 77-98, 11/22/77)
10. Indoor commercial recreational facilities such as motion picture theater, billiard parlor, swimming pool, bowling alley, and similar uses.
11. Vocational, trade, and business schools, provided all activities are conducted in completely enclosed buildings.
12. Miscellaneous uses such as telephone exchange, and commercial parking lots and parking garages.
13. Retail establishments manufacturing goods for sale only at retail on the premises.
14. Existing single family or two family dwellings.
15. Union hall.
16. Dry cleaning and laundry package plants in completely enclosed buildings using non-flammable liquids such as perchlorethylene and with no odor, fumes, or steam detectable to normal senses from off the premises.
17. Banks and financial institutions with or without drive-in facilities. (ORD. 81-52, 5/5/81)
18. Commercial and non-commercial piers, docks, etc., subject to the provisions of Sec. 7.23.
19. Railroad sidings.
20. Automotive service station, repair and service garage, motor vehicle body shop, where such facilities existed on the date of adoption of these regulations.
21. Animal hospital with boarding of animals in completely enclosed buildings.

Site and development plan approval (see Sec. 15.5) required for all uses on any lot or parcel within one hundred (100) feet of any residentially zoned property and, in addition, all uses are subject to the following limitations:

1. Sale, display, preparation, and repair incidental to sales, and storage to be conducted within a completely enclosed building.
2. Products to be sold only at retail. (ORD 78-95,10/3/78)

C. PERMITTED ACCESSORY USES AND STRUCTURES:

1. Uses and Structures which:
   a. Are customarily accessory and clearly incidental and subordinate to permitted or permissible uses and structures.
   b. Are located on the same lot as the permitted or permissible use or structure, or on a contiguous lot in the same ownership.
   c. Do not involve operations or structures not in keeping with the character of the district.
2. On the same premises and in connection with permitted principal uses and structures, dwelling units only for occupancy by owners or employees thereof.

D. PROHIBITED USES AND STRUCTURES:

1. Any uses or structures not specifically, provisionally, or by reasonable implication permitted herein.
   a. New single or two family dwellings.
   b. Manufacturing activities, except as specifically permitted or permissible.
   c. Warehousing or storage, except in connection with a permitted or permissible use.
   d. New adult entertainment establishments. (ORD. 77-103, 12/13/83)
2. Any use which is potentially dangerous, noxious, or offensive to neighboring uses in the district or to those who pass on public ways by reason of smoke, odor, noise, glare, fumes, gas, vibration, threat of fire or explosion, emission of particulate matter, interference with radio or television reception, radiation, or likely for other reason to be incompatible with the character of the district. Performance standards apply (see Sec. 9).

E. SPECIAL EXCEPTIONS:

(Permissible after Public Notice and Hearings by the Planning Commission and the Board of County Commissioners, see Section 20, "Special Exceptions.")

1. Package store for sale of alcoholic beverages, and bar or tavern for on-premises consumption of alcoholic beverages.
2. Automotive service station.
3. Public utility buildings and facilities necessary to serve surrounding neighborhoods (not including storage or service yards).
4. Multiple family dwellings (but not for one or two family dwellings).
5. Motor bus terminals.
6. Plant nursery.

7. Transient Accommodations (Principal Permitted and Accessory Uses as per "RTR" District, see S-60-a & b).
8. Boat livery.
9. Miniature golf-course.
10. Sale and display in other than completely enclosed building of any merchandise otherwise allowed as a permitted use in this district.
11. Buildings over thirty-five (35) feet in height but not in excess of eight-five (85) feet, provided an additional ten (10) feet for each story devoted primarily to parking within the structure up to a maximum additional height of twenty (20) feet may be added to the limit.
12. U.S. Post Office.
13. Emergency Services.
14. Night clubs (see Sec. 28.23), but not adult entertainment establishment.
15. Fast food restaurant.
16. Rehabilitative clinic. (ORD. 83-08, 2/15/83)

F. MAXIMUM RESIDENTIAL DENSITY:

(Dwelling units per acre, see Sec. 28.33, "Density, Residential" definition.)

1. Multiple Family Dwellings: Nine (9) units per acre.
2. Transient accommodations where not more than twenty-five percent (25%) of the units have cooking facilities:

Intensity Level Band (see Future Land Use Plan Map in *Apoxsee*)	Maximum Density (Subject to provisions of *Apoxsee*)
Band B	36
Band C	26
Band D	18
Band E	12

3. Transient accommodations where more than twenty-five percent (25%) of the units have cooking facilities:

Intensity Level Band (see Future Land Use Plan Map in *Apoxsee*)	Maximum Density (Subject to provisions of *Apoxsee*)
Band B	18
Band C	13
Band D	9
Band E	6

(ORD. 83-08, 2/15/83)

G. MINIMUM LOT REQUIREMENTS:

(Area and width, see Sec. 28.75, "Lot Measurement, Width" definition.)

1. Multiple family dwellings:
   a. Width: 100 feet
   b. Area: 2,420 sq. ft. per dwelling unit.
2. Other permitted or permissible uses and structures: None except as needed to meet other requirements herein set out.

H. MAXIMUM LOT COVERAGE BY ALL BUILDINGS:

(Includes accessory buildings, see Sec. 28.30, "Coverage of a Lot by Buildings" definition.)

1. Multiple family dwellings and their accessory buildings: Thirty percent (30%).
2. Other permitted or permissible buildings: Unrestricted, except as needed to meet other requirements herein set out.

I. MINIMUM YARD REQUIREMENTS:

(Depth of front and rear yards, width of side yards; see Sec. 28.141 - 28.146, "Yard" definitions; and see Sec. 7.14, "Base Setback Line Requirements.")

1. Commercial, service, office, or similar activities:
   a. Front:
      (1) Twenty (20) ft. if frontage of lot is one hundred (100) ft. or more, or if adjacent land is not built upon, or if buildings on adjacent lots have provided front yards of twenty (20) ft. or more.
      (2) If frontage is less than one hundred (100) ft. and if a building on an adjacent lot, or buildings on adjacent lots, provide yards less than twenty (20) ft. in depth, a front yard equal to average of adjacent front yards is required.
      (3) Where CG lot is located in a block or a portion of which is zoned residential, requirements of the residential district apply to said CG lot.
      (4) Not less than 1/3 of the depth of a front yard abutting a street shall be landscaped access: the remaining portion of the front yard may be used for offstreet parking but not for buildings.
   b. Side:
      (1) If fire resistive construction, building may be: (a) set to the side property line*, or (b) set not less than four (4) feet back from the side property line*.
      (2) If non-fire resistive construction, building must set back eight (8) feet from side property line. (County Building Codes should also be checked.)*

c. Rear: Ten (10) feet.*

*Special provisions: Where CG lot abuts property zoned residential, with or without an intervening alley, then at time of development of CG lot, landscaped buffer of the type specified in Section 7.22 (b) is required for rear and for side yards as the case may be. Such rear or side yards shall not be less than twenty (20) ft. in width or depth. No yards are required adjacent to mean high water line or adjacent to railroad tracks.

2. Multiple family dwellings:
   a. Front: *25 ft.
   b. Side: *15 ft.
   c. Rear: *15 ft.
   d. Waterfront: 20 ft. (see Section 7.11.c, "Minimum Gulf Beach Setback Line.")

   *Provided buildings above 35 ft. shall provide additional side and rear yards at a ratio of 1 ft. of yard for each 3 ft. of building height and front yard of 25 ft. or 1/2 of building height, whichever is greater.

3. Other permitted or permissible uses and structures, unless otherwise specified: Same as 1. above, "Commercial Service, Office, or Similar Activities."

J. MAXIMUM HEIGHT OF STRUCTURES:

(See Section 7.3, "Exclusions from Height Limits.")

All structures: 35 ft.

K. LIMITATIONS ON SIGNS:

(See also, generally, Section 14, "Signs" and Section 28.101-28.125 "Sign" definitions.)

1. No signs except:
   a. Exempt signs under Section 14.6(a).
   b. Business and professional offices, and libraries: One (1) ground or wall sign per street frontage, having no surface or facing exceeding sixteen (16) sq. ft. of signs mounted flush with the surface of a building; or having no single surface exceeding twelve (12) sq. ft. for hanging or ground signs, provided both sides of such hanging or freestanding signs may be used.
   c. Houses of worship: One (1) identification wall sign not exceeding twelve (12) sq. ft. in area and one (1) bulletin, ground, or wall sign not over twenty (20) sq. ft. in area for each street side.
   d. Private clubs: One (1) wall or ground sign, not over eight (8) sq. ft. in area.
   e. Multiple family dwellings: One (1) wall or ground sign, not to exceed thirty-two (32) sq. ft. in area, on each street frontage.

f. Wall, ground pylon, canopy, marquee, or projecting signs to advertise services or sale of products on the premises.
g. One (1) temporary construction project ground sign, not exceeding thirty-two (32) sq. ft. in area, such sign not to be erected more than sixty (60) days prior to time actual construction begins, and to be removed upon completion of actual construction. If construction is not begun within sixty (60) days or if construction is not continuously and actively prosecuted to completion, sign shall be removed.
h. Off-site signs (see Section 14.7 & 8).

2. General provisions:
   a. For signs in f. above, aggregate area of all signs shall not exceed three (3) sq. ft. in area for each foot of frontage occupied by building displaying signs, or one point five (1.5) sq. ft. in area for each foot of frontage on property occupied by building, whichever may be greater, provided no single business shall display more than two (2) signs, for each street frontage and provided no aggregate area of signs shall exceed two hundred (200) sq. ft. on a street frontage regardless of building or property frontage.
   b. No signs shall be erected in a manner than materially impedes visibility of moving vehicles or pedestrians on or off the premises: No sign (except projecting signs) shall be erected upon or overhang any street, right-of-way, walk, or alley except as specifically authorized.
   c. No ground sign shall be erected within fifty (50) ft. of any property zoned residential.
   d. Not more than one (1) sign structure may be erected in any required yard adjacent to a street, provided the area and number of signs on such structures shall be counted in the formula allocation of (a) above.

L. MINIMUM OFFSTREET PARKING REQUIREMENTS:

(See also, generally, Section 7.15, "Offstreet Vehicular Facilities - Parking and Loading.")

1. Commercial or service establishments (unless otherwise listed): 1 space for each 200 sq. ft. of non-storage floor area.
2. Professional or business office: 1 space for each 300 sq. ft. of non-storage floor area plus 1 space for each 2 occupants or employees.
3. Doctor and dentist office or clinic: 1 space for each doctor, nurse, and employee plus $1^1/_2$ spaces for each consultation and/or examining room.
4. Indoor motion picture theater: 1 space for each 3 seats.
5. Multiple family dwellings: 2 spaces for each dwelling unit plus two spaces for owner or manager, plus 2 spaces for each 3 employees.

6. Transient accommodations: 1 space for each sleeping room plus 1 additional space for each 10 sleeping rooms. Offstreet spaces shall not be used for storage of boats or boat trailers. (ORD 83-08, 2/15/83).
7. Restaurant, bar, or tavern: 1 space for each 3 seats in public rooms plus 1 space for each 2 employees.
8. House of worship: 1 space for each 3 seats in auditorium or chapel area.
9. Library or recreational facility: 1 space for each 200 sq. ft. of non-storage floor area or one (1) space for each 3 seats, whichever is greater.
10. Private clubs and lodges: 1 space for each 3 seats, or 1 space for each 300 sq. ft. of non-storage floor area, whichever is greater.
11. Funeral home: 1 space for each 2 seats in chapel.
12. Bowling alley: 6 spaces for each alley.
13. Billiard parlor: 1 space for each table.
14. Vocational, trade, or business schools: 1 space for each 250 sq. ft. of non-storage floor area or 1 space for each 4 seats, whichever is greater.
15. Miscellaneous uses such as telephone exchange: 1 space for each 2 employees plus 1 space for each 500 sq. ft. of floor area.
16. Marinas, commercial piers and docks: 2 spaces for each 3 boat slips or moorings.
17. Night club (but not adult entertainment establishment): 1 space for each 3 seats in public rooms plus 1 space for 2 employees, (ORD. 78-76, 7/18/78)
18. For other special exceptions, as specified herein: To be determined by general rule or by findings in the particular case.
19. Boat Liveries: 1 space for each 6 boats kept in dry storage. (ORD. 85-96, 8/6/85)

(Note: Accessory uses may require added spaces.)
(Note: Offstreet loading required.)

INDUSTRIAL GENERAL BUSINESS

(IG District)

OBJECTIVES

In order to promote the orderly growth and development of the community, protect the value of property, limit the expenditure of public funds, improve the opportunity for local employment and economic activity, and achieve the intent of land use regulations, this district is established to:

encourage the design and development of suitable areas for general types of industry which desire locations with direct access to major thoroughfares, railways, or other appropriate site selection criteria; and

discourage the creation or continuation of conditions which could detract from the function, operation, and appearance of industrial areas, or have an adverse effect upon adjacent areas.

PERMITTED USES

The following uses are authorized in this district, when such uses comply with the requirements contained in these regulations:

automobile service stations, garages for the overhaul and repair of motor vehicles, paint and body shops, tire recapping and car washing facilities;

shops and garages for overhaul and repair of engines, electric motors, air conditioners, heaters, radios, television sets, vacumn cleaners, vending machines, furniture or upholstery;

shops and garages for building trades such as plumbers, well drillers, electricians, cabinet builders, sheet metal workers, roofers, masons, tile setters, or general contractors;

laundries and plants for washing, drying and cleaning cloths, carpets, draperies, or household items;

areas for the display, sale, and maintenance of automobiles, trucks, travel trailers, motor bikes, boats, mobile homes, and farm or construction equipment;

areas for the display and sale of garden furniture and swimming pool equipment, or nurseries for propagation and sale of plants and landscaping materials;

areas for commercial recreation such as golf driving ranges, miniature golf, baseball and archery practice ranges;

drive-in restaurants and drive-in theaters;

warehouses and storage buildings.

CONDITIONAL USES

The following uses *may* be permitted as Conditional Uses provided that any review and hearing of an application for such a use shall consider the character of the area in which the proposed use is to be located, its effect on the value of surrounding lands, and the area of the site as it relates particularly to the required open spaces and off-street parking facilities.

Each application for a conditional use shall be accompanied by a site development plan which incorporates the regulations established here-in. The site development plan shall be drawn to scale and shall indicate property lines, rights-of-way, and the location of buildings, parking areas, curb cuts and driveways and open space strips. Said site development plan shall be reviewed by the Planning Commission and approved by the Board of County Commissioners prior to the granting of a land use and building permit. Upon such approval, said site development plan becomes part of the land use and building permit and may be amended only by the Board of County Commissioners.

plants for the production of concrete and masonry products such as ready-mix concrete, precast or prestressed structural members and building components, septic tanks, culvert or pipe;

plants for the fabrication of wood products such as roof trusses, modular units and cabinets including the processing of logs or stumps and the sizing or curing of lumber;

plants for the fabrication of steel products such as roof trusses, bar joists and decorative panels, including machine shops and related tool or die work;

plants for the production of plastic, glass or ceramic products such as pipe, hose, molding strips, blocks or pottery;

plants for the processing of food products such as citrus, fruits, vegetables, poultry or cattle;

kennels and animal clinics for services customarily provided by veterinarians, and auction facilities for livestock;

plants for the production of asphaltic materials provided that such plants meet all Federal and State agencies' noise and emission control requirements; (ZTA 85-01 3/18/85)

tank farms for bulk storage of petroleum and other liquids;

storage and docking facilities for repair and maintenance of boats and marine craft;

terminals for trucking;

plants for the generating of electricity and the treating of water or sewage;

yards for storage of materials, resources, products and equipment, including the salvage and reclamation of used items such as metal, paper, lumber, cloth or other scrap materials;

other similar uses which are reasonably implied and are consistent with the *objectives* of this district, based on appropriate consideration of the *nature* of the intended activity, the *character* of the proposed development, the *location* of the site, and its *compatibility* with adjacent parcels.

PROHIBITED USES

All uses not specified as a Permitted Use or approved as a Conditional Use shall be prohibited. The following are examples of uses which are not authorized in this district:

plants for the production or batching of asphalt or asphaltic concrete;

mining or quarry operations;

reclamation or rendering of animal fats or wastes;

plants for the processing of fertilizer or animal hides;

other similar uses which are not compatible with the objectives and intent of this district, or activities which would not be beneficial to the community by endangering the public, polluting the environment, unbalancing the ecology, or otherwise lowering the quality of life for its citizens.

DEVELOPMENT STANDARDS

Minimum Lot Area 20,000 square feet
Minimum Lot Width 150 feet
Maximum Building Coverage 50 percent of lot area
Maximum Building Height 4 stories
Minimum Building Setbacks

The following setbacks are measured from the existing property line:

front yard	35 feet (NOTE 1)
rear yard	25 feet
side yard	15 feet (NOTE 2) (NOTE 1)

In addition, one of the setbacks specified below (measured from the centerline of adjacent public rights-of-way) shall apply if *greater* than the setback specified above from the property line.

	Planned Right-of-Way		Setback From Centerline
(A)	300 feet maximum right-of way (NOTE 3) 50 feet minimum right-of-way	(A)	one half proposed right-of-way plus required yard (front, rear, or side)
(B)	thoroughfares designated on Road Plans as having a width of 100 feet or more	(B)	one half proposed right-of-way plus 50 feet
(C)	thoroughfares having a width of less than 100 feet	(C)	one half proposed right-of-way plus 25 feet

Minimum Off-Street Parking/Loading (See separate section of regulations)

Paving

All on-site parking spaces and all on-site driveways shall be paved.

Landscaping

A five (5) foot open space strip, suitable for landscaping, shall be retained on all sides of the property. Existing natural resources such as tree stands or large individual trees, will be conserved whenever practical to do so. Private signs, lights, decorative features, and other similar items will be removed at no expense to the public from such open space strips if these areas should subsequently be required for the construction of service drives, access roads, or other public improvements.

NOTE 1: Automobile service stations (and other uses having a similar relationship with motor vehicles) shall locate the structure in such a manner that the pump island will be set back an additional ten (10) feet to allow for servicing of the automobile within the authorized building site—not in the required open space between the setback line and the road right-of-way.

NOTE 2: The minimum setback specified above for side yards shall apply to the first two stories only. This minimum side-yard setback may be decreased by not more than five (5) feet on one side (provided the setback on the other side yard is increased by an equal distance); however, the minimum side-yard setback may not be decreased for any side yard which is adjacent to a public roadway. For buildings that are three or more stories in height, the minimum side-yard setback for that portion of the building above the first two stories shall be computed as follows:

the maximum number of stories in the
building multiplied by ten (10) feet.

NOTE 3: Maximum right-of-way may be greater at interchanges, intersections, or in other locations where additional right-of-way is required for public purposes.

COMMERCIAL RESTRICTED BUSINESS

(CR District)

OBJECTIVES

In order to promote the orderly growth and development of the community, protect the value of property, limit the expenditure of public funds, improve the opportunity for local employment and economic activity, and achieve the intent of land use regulations, this district is established to:

encourage the design and development of suitable areas for special types of commerce which desire locations with direct access to major thoroughfares and offer a variety of goods and services to various regions of the community; and

discourage the creation or continuation of conditions which could detract from the function, operation, and appearance of planned business centers, or have an adverse effect on adjacent areas.

CONDITIONAL USES

The following uses *may* be permitted as Conditional Uses provided that any review and hearing of an application for such a use shall consider the character of the area in which the proposed use is to be located, its effect on the value of surrounding lands, and the area of the site as it relates particularly to the required open spaces and off-street parking facilities.

Each application for a conditional use shall be accompanied by a site development plan which incorporates the regulations established herein. The site development plan shall be drawn to scale and shall indicate property lines, rights-of-way, and the location of buildings, parking areas, curb cuts and driveways. Said site development plan shall be reviewed by the Planning Commission and approved by the Board of County Commissioners prior to the granting of a land use and building permit. Upon such approval, said site development plan becomes part of the land use and building permit and may be amended only by the Board of County Commissioners.

shopping centers and office parks;

shops and stores for retail goods such as furniture, appliances, clothing, food, medicine, hardware, books, gifts, carpets, draperies, stationery, office supplies, or sporting goods;

shops and studios for general services such as hair cutting and styling, shoe repair, photography, picture framing, and areas for pick-up and delivery of laundry or cleaning;

offices and studios for professional services such as customarily provided by doctors, dentists, opticians, architects, engineers, or lawyers;

offices and studios for financial services such as customarily provided by banks, savings and loan associations, credit bureaus, insurance agencies, and brokers for real estate or securities;

offices and studios for communication services such as radio and television broadcasting, filming or recording, and publishing of newspapers or periodicals;

offices and studios for business services such as data processing, employee training, advertising, bookkeeping, and duplicating or reproducing of letters, forms, and drawings;

administrative buildings and courts for governmental services such as executive, legislative, and judicial functions of local, state, and federal agencies, including post office facilities;

religious facilities such as churches, chapels, and educational buildings;

eating establishments such as restaurants, cafeterias, lounges, supper clubs, and coffee shops;

theaters, museums, and galleries for cultural events such as plays, motion pictures, displays and art shows; and clubs, lodges, or meeting facilities for civic, professional or social organizations;

entertainment facilities such as auditoriums, bowling alleys, skating rinks, and dance halls;

hospitals, medical clinics, or laboratories and customary accessory uses;

warehouses for storage and distribution of supplies used in conjunction with goods or services authorized in this district;

transportation facilities such as passenger stations for trains, buses, limousines, or taxis, and parking garages for automobiles;

food preparation facilities for minor products to be sold on the premises, such as bakery items, candies, confections, juices and beverages;

production facilities for limited communication products such as newspapers, magazines, and similar printing, lithography or photographic processing;

automobile service stations;

showrooms for new cars and customary accessory uses;

other similar uses which are reasonably implied and are consistent with the *objectives* of this district, based on appropriate consideration of the *nature* of the intended activity, the *character* of the proposed development, the *location* of the site, and the *compatibility* with adjacent parcels.

PROHIBITED USES

All uses not specified as a Permitted Use or approved as a Conditional Use shall be prohibited. The following are examples of uses which are not authorized in this district:

drive-in restaurants and drive-in theaters;

animal clinics for services customarily provided by veterinarians;

used car lots;

other similar uses which are not compatible with the objectives and intent of this district, or activities which would not be beneficial to the community by endangering the public, polluting the environment, unbalancing the ecology, or otherwise lowering the quality of life for its citizens.

DEVELOPMENT STANDARDS

Minimum Lot Area	20,000 square feet
Minimum Lot Width	150 feet
Maximum Building Coverage	50 percent of lot area
Maximum Building Height	4 stories

Minimum Building Setbacks

The following setbacks are measured from the existing property line:

front yard	35 feet (NOTE 1)
rear yard	25 feet
side yard	15 feet (NOTE 2) (NOTE 1)

In addition, one of the setbacks specified below (measured from the centerline of adjacent public rights-of-way) shall apply if *greater* than the setback specified above from the property line.

Planned Right-of-Way	Setback From Centerline
(A) 300 feet maximum right-of-way (NOTE 3)	(A) one half proposed right-of-way plus required yard (front, rear, or side)
50 feet minimum right-of way	
(B) thoroughfares designated on Road Plans as having a width of 100 feet or more	(B) one half proposed right-of-way plus 50 feet
(C) thoroughfares having a width of less than 100 feet	(C) one half proposed right-of-way plus 25 feet

Minimum Off-Street Parking/Loading (See separate section of regulations)

REQUIRED STATEMENTS

The site development plan, which is required in conjunction with all requests for Conditional Uses, shall indicate areas for and the extent of landscaping to be provided, and shall include a statement of intent similar to the following:

Landscaping

> The land will not be overdeveloped with buildings, structures, or paving. Open spaces will be an integral part of the site and landscaping will be provided and maintained to enhance the character and appearance of the development. Existing natural resources, such as tree stands or large individual trees, will be conserved, whenever practical to do so.

The site development plan shall also illustrate setback lines and shall contain a statement of intent similar to the following:

Setbacks

> Necessary removal of private signs, lights, decorative features, and other similar items from open spaces between road rights-of-way and building setback lines will be accomplished at no expense to the public if these areas should subsequently become required for the construction of service drives, access roads, or other public improvements.

NOTE 1: Automobile service stations (and other uses having a similar relationship with motor vehicles) shall locate the structure in such a manner that the pump island will be set back an additional ten (10) feet to allow for servicing of the automobile within the authorized building site—not in the required open space between the setback line and road right-of-way.

NOTE 2: The minimum setback specified above for side yards shall apply to the first two stories only. This minimum side-yard setback may be decreased by not more than five (5) feet on one side (provided the setback on the other side yard is increased by an equal distance); however, the minimum side-yard setback may not be decreased for any side yard which is adjacent to a public roadway. For buildings that are three or more stories in height, the minimum side-yard setback for that portion of the building above the first two stories shall be computed as follows:

the maximum number of stories in the
building multiplied by ten (10) feet.

NOTE 3: Maximum right-of-way may be greater at interchanges, intersections, or other locations where additional right-of-way is required for public purposes.

RESIDENTIAL MULTIPLE FAMILY

(RM-1 District)

OBJECTIVES

In order to promote the orderly growth and development of the community, protect the value of property, improve the opportunity for housing various economic and other groups, and achieve the intent of land use regulations, this district is established to:

encourage the design and development of suitable areas for various types of residential dwellings at a medium density—such as detached low cost homes for single family use and attached units for multiple family use; and

discourage the creation or continuation of conditions which could detract from the harmony, tranquillity and appearance of residential neighborhoods, or have an adverse effect on adjacent areas.

PERMITTED USES

The following uses are authorized in this district, when such uses comply with the requirements contained in these regulations:

single family dwelling (*house*), including customary uses such as attached garage, carport and porches, when accessory and incidental to a single family house;

swimming pool, bathhouse, and customary uses, when accessory and incidental to a single family house (or duplex), provided the pool is not located within the required front yard, or closer than eight (8) feet to any side or rear property line, or within any required side yard adjacent to a public roadway;

greenhouse or slathouse for domestic plants and landscaping materials, when accessory and incidental to a single family house (or duplex), provided such structures are approved by the Board of Adjustment as a Special Exception;

stable for horses, when accessory and incidental to a single family house, provided such structures are approved by the Board of Adjustment as a Special Exception, and provided the density does not exceed one horse/two acres of pasture;

pens for dogs and other domestic pets, when accessory and incidental to a single family house (or duplex), provided such structures are approved by the Board of Adjustment as a Special Exception;

detached garage, hobby shop, storage building, and tool shed, when accessory and incidental to a single family house (or duplex);

pier, dock, or boathouse, when accessory and incidental to a single family house (or duplex), provided such structures are not located closer to any side line (extended) than the specified setback for side yards.

CONDITIONAL USES

The following uses *may* be permitted as Conditional Uses provided that any review and hearing of an application for such a use shall consider the character of the area in which the proposed use is to be located, its effect on the value of surrounding lands, and the area of the site as it relates particularly to the required open spaces and off-street parking facilities.

Each application for a conditional use shall be accompanied by a site development plan which incorporates the regulations established herein. The site development plan shall be drawn to scale and shall indicate property lines, rights-of-way, and the location of buildings, parking areas, curb cuts and driveways. Said site development plan shall be reviewed by the Planning Commission and approved by the Board of County Commissioners prior to the granting of a land use and building permit. Upon such approval, said site development plan becomes part of the land use and building permit and may be amended only by the Board of County Commissioners.

two family dwelling *(duplex)*, including customary uses such as attached garage, carport and porches, when accessory and incidental to a duplex;

parks, playgrounds, libraries, and similar neighborhood activities not operated for profit;

sub-stations for telephone, electric power, or other utilities, and for fire fighting or law enforcement services;

churches and customary accessory facilities such as chapels and educational buildings for religious training;

kindergartens and child care centers for pre-school children;

schools and customary accessory facilities such as cafeterias, auditoriums, gymnasiums, and ball fields;

neighborhood recreation facilities not operated for profit such as clubhouses, swimming pools, picnic grounds, beaches, bathhouses, boat docks, and boat ramps;

marinas or golf courses, country clubs, and customary accessory facilities such as clubhouses, swimming pools, cabanas, tennis courts, maintenance buildings, and structures for storage of golf carts;

other similar uses which are reasonably implied and are consistent with the *objectives* of this district, based on appropriate consideration of the *nature* of the intended activity, the *character* of the proposed development, and *location* of the site, and its *compatibility* with adjacent parcels.

PROHIBITED USES

All uses not specified as a Permitted Use or approved as a Conditional Use shall be prohibited. The following are examples of uses which are not authorized in this district:

commercial enterprises operated for profit such as, guest houses, plant nurseries, stables, kennels, auditoriums, or stadiums;

mobile homes and similar forms of manufactured housing;

other similar uses which are not compatible with the objectives and intent of this district, or activities which would not be beneficial to the community by endangering the public, polluting the environment, unbalancing the ecology, or otherwise lowering the quality of life for its citizens.

DEVELOPMENT STANDARDS

Minimum Lot Area

house	7,000 square feet
duplex	9,500 square feet

Minimum Lot Width

house	70 feet
duplex	95 feet

Maximum Building Coverage 50 percent of lot area

Maximum Building Height 2 stories

Maximum Dwelling Density

house	5.0 units/acre
duplex	7.3 units/acre

Minimum Living Space 750 square feet

Minimum Building Setbacks

The following setbacks are measured from the existing property line:

front yard	25 feet
rear yard	25 feet
side yard	10 feet

In addition, one of the setbacks specified below (measured from the centerline of adjacent public rights-of-way) shall apply if *greater* than the setback specified above from the property line.

Planned Right-of-Way		Setback From Centerline	
(A)	300 feet maximum right-of-way (NOTE)	(A)	one half proposed right-of-way plus required yard (front, rear, or side)
	50 feet minimum right-of-way		
(B)	thoroughfares designated on Road Plans as having a width of 100 feet or more	(B)	one half proposed right-of-way plus 50 feet
(C)	thoroughfares having a width of less than 100 feet	(C)	one half proposed right-of-way plus 25 feet

Minimum Off-Street Parking

dwellings	2 spaces/dwelling unit
all other requirements (parking, loading, ingress and egress)	(See separate section of regulations)

REQUIRED STATEMENTS

The site development plan, which is required in conjunction with all requests for Conditional Uses, shall indicate areas for and the extent of landscaping to be provided, and shall include a statement of intent similar to the following:

Landscaping

The land will not be overdeveloped with buildings, structures, or paving. Open spaces will be an integral part of the site and landscaping will be provided and maintained to enhance the character and appearance of the development. Existing natural resources, such as tree stands or large individual trees, will be conserved, whenever practical to do so.

The site development plan shall also illustrate setback lines and shall contain a statement of intent similar to the following:

Setbacks

Necessary removal of private signs, lights, decorative features, and other similar items from open spaces between road rights-of-way and building setback lines will be accomplished at no expense to the public if these areas should subsequently become required for the construction of service drives, access roads, or other public improvements.

NOTE 1: Maximum right-of-way may be greater at interchanges, intersections, or in other similar locations where additional right-of-way is required for public purposes.

NOTE 2: Congregate Living Facilities for unlimited number of people may be allowed as a Conditional Use in RM-1, RM-2 and RM-3 Districts provided that internal facilities have been approved by HRS.

RESIDENTIAL SINGLE FAMILY

(RS-1 – RS-3 Districts

and RS-1A and RS-1C Districts)

OBJECTIVES

In order to promote the orderly growth and development of the community, protect the value of property, improve the opportunity for housing various economic and other groups, and achieve the intent of land use regulations, this district is established to:

encourage the design and development of suitable areas for select types of residential dwellings at a low density—such as detached, high and medium value homes for single family use; and

discourage the creation or continuation of conditions which could detract from the harmony, tranquillity, and appearance of residential neighborhoods, or have an adverse effect on adjacent areas.

PERMITTED USES

The following uses are authorized in this district, when such uses comply with the requirements contained in these regulations:

single family dwelling (*house*), including customary uses such as attached garage, carport and porches, when accessory and incidental to a single family house;

swimming pool, bathhouse, and customary uses, when accessory and incidental to a single family house, provided the pool is not located within the required front yard, or closer than eight (8) feet to any side or rear property line, or within any required side yard adjacent to a public roadway;

guest cottage or quarters for domestic employees (on lots having a minimum area of 25,000 square feet), when accessory and incidental to a single family house, provided such structures are approved by the Board of Adjustment as a Special Exception;

greenhouse or slathouse for domestic plants and landscaping materials, when accessory and incidental to a single family house, provided such structures are approved by the Board of Adjustment as a Special Exception;

stable for horses (on lots of two acres or more), when accessory and incidental to a single family house, provided such structures are approved by the Board of Adjustment as a Special Exception;

pens for dogs and other domestic pets, when accessory and incidental to a single family house, provided such structures are approved by the Board of Adjustment as a Special Exception;

detached garage, hobby shop, storage building, or tool shed, when accessory and incidental to a single family house;

pier, dock, or boathouse, when accessory and incidental to a single family house, provided such structures are not located closer to any side lot line (extended) than the specified setback for side yards.

Congregate Living Facilities for no more than five (5) individuals provided that internal facilities have been approved by HRS; provided a manager lives on site.

CONDITIONAL USES

The following uses *may* be permitted as Conditional Uses provided that any review and hearing of an application for such a use shall consider the character of the area in which the proposed use is to be located, its effect on the value of surrounding lands, and the area of the site as it relates particularly to the required open spaces and off-street parking facilities.

Each application for a conditional use shall be accompanied by a site development plan which incorporates the regulations established here-in. The site development plan shall be drawn to scale and shall indicate property lines, rights-of-way, and the location of buildings, parking areas, curb cuts and driveways. Said site development plan shall be reviewed by the Planning Commission and approved by the Board of County Commissioners prior to the granting of a land use and building permit. Upon such approval, said site development plan becomes part of the land use and building permit and may be amended only by the Board of County Commissioners.

parks, playgrounds, libraries, and similar neighborhood activities not operated for profit;

sub-stations for telephone, electric power, or other utilities, and for fire fighting or law enforcement services;

churches and customary accessory facilities such as chapels and educational buildings for religious training;

kindergartens and child care centers for pre-school children;

schools and customary accessory facilities such as cafeterias, auditoriums, gymnasiums, and ball fields;

neighborhood recreation facilities not operated for profit such as clubhouses, swimming pools, picnic grounds, beaches, bathhouses, boat docks, and boat ramps;

marinas or golf courses, country clubs, and customary accessory facilities such as clubhouses, swimming pools, cabanas, tennis courts, maintenance buildings, and structures for storage of golf carts;

Congregate Living Facilities for more than five (5) individuals but less than ten (10) may be a Conditional Use provided that internal facilities have been approved by HRS;

other similar uses which are reasonably implied and are consistent with the *objectives* of this district, based on appropriate consideration of the *nature* of the intended activity, the *character* of the proposed development, the *location* of the site, and its *compatibility* with adjacent parcels.

PROHIBITED USES

All uses not specified as a Permitted Use or approved as a Conditional Use shall be prohibited. The following are examples of uses which are not authorized in this district:

commercial enterprises operated for profit such as guest houses, plant nurseries, stables, kennels, auditoriums or stadiums;

mobile homes and similar forms of manufactured housing;

other similar uses which are not compatible with the objectives and intent of this district, or activities which would not be beneficial to the community by endangering the public, polluting the environment, unbalancing the ecology, or otherwise lowering the quality of life for its citizens.

DEVELOPMENT STANDARDS

Minimum Lot Area

RS-1 District	15,000 square feet/dwelling unit
RS-1A District	15,000 square feet/dwelling unit
RS-1C District	21,780 square feet/dwelling unit
RS-2 District	10,000 square feet/dwelling unit
RS-3 District	7,500 square feet/dwelling unit

Minimum Lot Width

RS-1 District	100 feet
RS-1A District	100 feet
RS-1C District	100 feet
RS-2 District	85 feet
RS-3 District	75 feet

Maximum Building Coverage 50 percent of lot area

Maximum Building Height 2 stories

Maximum Dwelling Density

RS-1 District	2.5 units/acre
RS-1A District	2.5 units/acre
RS-1C District	2.0 units/acre
RS-2 District	3.6 units/acre
RS-3 District	4.6 units/acre

Minimum Living Space

RS-1 District	1,500 square feet/dwelling unit
RS-1A District	1,200 square feet/dwelling unit
RS-1C District	1,000 square feet/dwelling unit
RS-2 District	1,000 square feet/dwelling unit
RS-3 District	750 square feet/dwelling unit

Minimum Building Setbacks

The following setbacks are measured from the existing property line:

	RS-1	RS-1A & C	RS-2	RS-3
front yard	25 feet	25 feet	25 feet	25 feet
rear yard	25 feet	25 feet	25 feet	25 feet
side yard	15 feet	15 feet	12 feet	10 feet

In addition, one of the setbacks specified below (measured from the centerline of adjacent public rights-of-way) shall apply if *greater* than the setback specified above from the property line.

Planned Right-of-Way	Setback From Centerline
(A) 300 feet maximum right-of-way (NOTE 1)	(A) one half proposed right-of-way plus required yard (front, rear, or side)
50 feet minimum right-of way	
(B) thoroughfares designated on Road Plans as having a width of 100 feet or more	(B) one half proposed right-of-way plus 50 feet
(C) thoroughfares having a width of less than 100 feet	(C) one half proposed right-of-way plus 25 feet

Minimum Off-Street Parking

dwellings	2 spaces/dwelling unit
all other requirements (parking, loading, ingress and egress)	(See separate section of regulations)

REQUIRED STATEMENTS

The site development plan, which is required in conjunction with all requests for Conditional Uses, shall indicate areas for and the extent of landscaping to be provided, and shall include a statement of intent similar to the following:

Landscaping

The land will not be overdeveloped with buildings, structures, or paving. Open spaces will be an integral part of the site and landscaping will be provided and maintained to enhance the character and appearance of the development. Existing natural resources, such as tree stands or large individual trees, will be conserved, whenever practical to do so.

The site development plan shall also illustrate setback lines and shall contain a statement of intent similar to the following:

Setbacks

Necessary removal of private signs, lights, decorative features, and other similar items from open spaces between road rights-of-way and building setback lines will be accomplished at no expense to the public if these areas should subsequently become required for the construction of service drives, access roads, or other public improvements.

NOTE 1: Maximum right-of-way may be greater at interchanges, intersections, or in other similar locations where additional right-of-way is required for public purposes.
NOTE 2: RS-1C to be allowed in the Community Development Areas ONLY.

COMMERCIAL TOURIST CENTER

(CT District)

OBJECTIVES

In order to promote the orderly growth and development of the community, protect the value of property, limit the expenditure of public funds, improve the opportunity for local employment and economic activity, and achieve the intent of land use regulations, this district is established to:

encourage the design and development of suitable areas for special types of commerce which desire locations with direct access to major thoroughfares and offer a variety of goods and services primarily for the tourist market; and

discourage the creation or continuation of conditions which could detract from the function, operation, and appearance of planned tourist centers, or have an adverse effect on adjacent areas.

CONDITIONAL USES

The following uses *may* be permitted as Conditional Uses provided that any review and hearing of an application for such a use shall consider the character of the area in which the proposed use is to be located, its effect on the value of surrounding lands, and the area of the site as it relates particularly to the required open spaces and off-street parking facilities.

Each application for a conditional use shall be accompanied by a site development plan which incorporates the regulations established herein. The site development plan shall be drawn to scale and shall indicate property lines, rights-of-way, and the location of buildings, parking areas, curb cuts and driveways. Said site development plan shall be reviewed by the Planning Commission and approved by the Board of County Commissioners prior to the granting of a land use and building permit. Upon such approval, said site development plan becomes part of the land use and building permit and may be amended only by the Board of County Commissioners.

transient housing facilities (hotel and motel complexes), including customary uses such as garages, carports, recreation rooms, swimming pools, bathhouses, piers, docks or boathouses, storage or maintenance buildings, and areas providing convenience goods or services for occupants, when designed as an integral and compatible unit of the complex, and when accessory and incidental to the hotel or motel;

convention facilities such as auditoriums, conference rooms or display areas;

eating establishments such as restaurants, cafeterias, lounges, supper clubs, coffee shops, or stores for the preparation of fast-food items and similar take-out orders;

shops and stores for convenience goods such as gifts, souvenirs, clothing, flowers, books, artists supplies, craft or hobbies;

shops and stores for convenience services such as cutting and styling of hair, self-service facilities for washing, drying and cleaning clothes, or areas for pick-up and delivery of laundry and cleaning;

facilities for recreation such as bowling alleys, theater buildings, golf driving ranges, or putting practice areas;

tourist attractions such as glass blowing exhibits, wax museum, or museum for antique cars; shops for the fabrication of arts, crafts or souvenirs; facilities for the preparation, display and sale of candies, jam, preserves, honey, or other agricultural products;

automobile service stations, provided that repair work is limited to minor auto repair (not the overhaul of engines, transmissions or rear ends, and not welding, body work, painting, upholstery, or tire recapping);

time sharing units to provide transient housing; these units will not be leased on a nightly basis; no unit shall be sold to an individual or group for a time period greater than eight weeks; and the management group will be responsible for collecting the taxes for all units; no motel conversions will be permitted; all condominium documents must be approved by the County Attorney;

other similar uses which are reasonably implied and are consistent with the *objectives* of this district, based on the appropriate consideration of the *nature* of the intended activity, the *character* of the proposed development, and *location* of the site, and its *compatibility* with adjacent parcels.

PROHIBITED USES

All uses not specified as a Permitted Use or approved as a Conditional Use shall be prohibited. The following are examples of uses which are not authorized in this district:

travel trailer parks and mobile home parks;

drive-in restaurant, except that drive-in services are authorized in conjunction with a conventional restaurant, provided the service is accessory and incidental to a restaurant that has at least two customer-seats within the building for every automobile space provided with drive-in type service, and provided drive-in facilities are designed as an integral and compatible unit of the complex;

major laundries and plants for washing, drying, and cleaning clothes, carpets, draperies, or other household items;

drive-in theaters;

truck terminals and garages for the overhaul or repair of motor vehicles;

areas for the sale, display and rental of mobile homes, travel trailers, automobiles, trucks, boats, or other miscellaneous trailers, except that an auto rental service is authorized in conjunction with a hotel or motel to serve their customers, provided the service is accessory and incidental to the principal use:

areas for the sale of agricultural products, candies and souvenirs, except that these uses are authorized when housed within an enclosed building that meets the requirements of all construction codes;

temporary and other transient commercial recreation facilities such as carnivals, super slides, trampolines, and tracks or drag strips for autos, motorcycles, or go-carts;

uses which require outside storage of materials or equipment, unless completely screened by an approved wall or landscaped buffer;

the manufacture, fabrication, processing and assembly of goods or merchandise other than those authorized as conditional use:

other similar uses which are not compatible with the objectives and intent of this district, and activities which would not be beneficial to the community by endangering the public, polluting the environment, unbalancing the ecology, or otherwise lowering the quality of life for its citizens.

DEVELOPMENT STANDARDS

Minimum Lot Area 20,000 square feet
Minimum Lot Width 150 feet
Maximum Building Coverage 50 percent of lot area
Maximum Building Height No Limit (Amended 12/3/79 ZTA 79-03) NOTE 4

Maximum Unit Density
hotel-motel complex 45 units per acre
time sharing units 14 units per acre

Minimum Building Setbacks (ZTA 82-04 5/17/82)

The following setbacks are measured from the existing property line:

front yard 35 feet (NOTE 1)

rear yard	25 feet
side yard	15 feet (NOTE 1) (NOTE 2)

In addition, one of the setbacks specified below (measured from the centerline of adjacent public rights-of-way) shall apply if *greater* than the setback specified above from the property line.

Planned Right-of-Way	**Setback From Centerline**
(A) 300 feet maximum right-of-way (NOTE 3) 50 feet minimum right-of way	(A) one half proposed right-of-way plus required yard (front, rear, or side)
(B) thoroughfares designated on Road Plans as having a width of 100 feet or more	(B) one half proposed right-of-way plus 50 feet
(C) thoroughfares having a width of less than 100 feet	(C) one half proposed right-of-way plus 25 feet

Minimum Off-Street Parking/Loading (See separate section of regulations)

time sharing units	1.5 spaces per unit (ZTA 82-04 5/17/82)

Minimum Living Area

time sharing units	600 square feet per unit (ZTA 82-04 5/17/82)

REQUIRED STATEMENTS

The site development plan, which is required in conjunction with all requests for Conditional Uses, shall indicate areas for and the extent of landscaping to be provided, and shall include a statement of intent similar to the following:

Landscaping

The land will not be overdeveloped with buildings, structures, or paving. Open spaces will be an integral part of the site and landscaping will be provided and maintained to enhance the character and appearance of the development. Existing natural resources, such as tree stands or large individual trees, will be conserved, whenever practical to do so.

The site development plan shall also illustrate setback lines and shall contain a statement of intent similar to the following:

Necessary removal of private signs, lights, decorative features, and other similar items from open spaces between road rights-of-way and building setback lines will be accomplished at no expense to the public if these areas should subsequently become required for the construction of service drives, access roads, or other public improvements.

NOTE 1: Automobile service stations (and other uses having a similar relationship with motor vehicles) shall locate the structure in such a manner that the pump island will be set back an additional ten (10) feet to allow for servicing of the automobile within the authorized building site—not in the required open space between the setback line and the road right-of-way.
NOTE 2: The minimum setback specified above for side yards shall apply to the first four stories only.
NOTE 3: Maximum right-of-way may be greater at interchanges, intersections, or in other locations where additional right-of-way is required for public purposes.
NOTE 4: Buildings above four stories will be allowed providing they are in compliance with the following standards:

1. Three feet additional setback shall be provided for each story over 4.
2. Shall not conflict with the air traffic zones of the Kissimmee and Orlando Airports.
3. For safety and fire protection purposes shall be built in compliance with the Southern Building Code for high rise structures. (Amended ZTA 79-03 December 3, 1979)

Glossary

abstract of title—a summary of all transactions in the continuing ownership of property.

acceleration lane—the paved strip of highway you pull onto to assist you in entering the flow of traffic by first building speed.

access—the right to enter and leave a tract of land usually from a public road or street.

ADI—see area of dominant influence.

adjustable rate mortgage—a mortgage that has an interest that fluctuates according to a predetermined schedule.

amortize—this term is meant to say that you will pay your loan over a specific period of time.

amortization—the periodic payment of a loan.

appraisal—an opinion of value arrived at by comparing property sizes and characteristics. Also see M.A.I.

area of dominant influence (ADI)—a group of counties in which the radio and/or television stations get the largest audience share. Also see DMA.

architect—a person whose profession is designing, drawing, and supervising the construction of buildings, etc.

assessment—a charge added to your real estate taxes because of a service or improvement made from which you would benefit, i.e., if the dirt road in front of your property is paved, a proportionate cost could be added to your real estate tax.

attorney—a person trained in the law to advise or represent others in legal matters.

barrier—an obstruction or something that prohibits or hinders passage. This could be a natural barrier such as a mountain or river, or a man-made barrier such as an interstate highway or a bridge.

bedroom community—an area of residential development. It is called a bedroom community because it is located conveniently to, but out of, the work area. This concept is particularly popular in the larger, and especially, industrial cities.

blanket zoning—zoning enacted to cover a group of properties according to the comprehensive land use plan. Usually used for uniformity in road maintenance, police and fire services, etc.

bottom line—a term used in relation to the feasibility study to mean the profit after all expenses are paid, and all taxes are addressed.

boundary lines—the outline of your property. An easterly boundary line shows the end (or beginning) of your property to the east, etc.

builder's risk insurance—insurance for fire, windstorm, lightning etc. for a building under construction.

building codes—governmental regulations specifying minimum construction standards. These codes limit the use, size, and location of the buildings on the property.

CCIM—(certified commercial investment member.) A designation awarded by the Realtor's National Marketing Institute to a realtor or realtor-associate active in the field of commercial real estate who has achieved a superior level of knowledge by completing the requirements set out by the Realtor's National Marketing Institute.

CPUD—(commercial planned unit development.) A piece of property platted as zoned according to a particular development plan.

CRS—(certified residential specialist.) A designation awarded by the Realtor's National Marketing Institute to a realtor or realtor-associate active in the field of residential real estate who has achieved a superior level of knowledge by completing the requirements set out by the Realtor's National Marketing Institute.

city or city limits—an incorporated municipality whose boundaries and powers are defined by a charter.

city taxes—the tax on real estate, sales, salaries, etc. imposed by a city.

city utilities—the utility services that are supplied and maintained by the city.

civil engineer—an engineer who specializes in the planning and building of roads and bridges.

clause—a statement in a written document (contract) that usually outlines a stipulation, restriction, etc.

commercial impact area—an area that, because of development, provides the population with a reason to stop there. Especially an area with more than one place to shop or entertain.

comprehensive land use plan—an overall plan usually set up by the county or region to control the zoning and development of the county or region.

contingency—a clause in the contract that bases the purchase on a future condition. For instance, the contract will close on a certain date, contingent upon receipt of a building permit.

contractor—a person trained and licensed in a particular field of expertise, such as road paving or building construction.

counteroffer—an offer made in response to an offer that is not acceptable.

county—the largest local administrative subdivision of most states in which cities are formed.

county tax—any payroll, real estate or other tax required by the county in which you are working.

cross easement—(Also see easement.) The right to go onto someone else's property for a specified reason i.e., a cross easement for parking gives you the right to park on the property as specified.

cross parking—the right to park on someone else's property as agreed upon, usually by contract.

curb cut—this term is meant to identify the entrance to your property from a road or highway granted by the department of transportation.

customer service business—any business that provides a service to a customer, whether directly or indirectly. Every business in one way or another is a customer service business.

deceleration lane—a paved strip of highway off the main road that the motorists pull onto to slow down before turning a corner or into an establishment.

deed restriction—a restriction put on a deed to limit a specific use or uses.

demographics—statistics regarding the population, occupation, income level, age, race, etc. of an area or a town.

demuck—the removal of muck, involving digging out the muck and replacing it with good soil.

density—the number of apartments or houses you are allowed by zoning per acre.

department of environmental protection—(Also may be referred to as the department of environmental resources.) A government agency set up to protect the environment.

department of environmental resources—see department of environmental protection.

department of natural resources—a government agency set up to protect our natural resources, such as water and air.

directive of regional impact (DRI)—A study that is usually required for projects over 100 acres that studies how the environment in its entirety will be affected by a project.

department of transportation—the department set up on a state and national level to plan, design, and control traffic.

designated market area (DMA)—(Also see area of dominate influence.) A group of counties in which the radio and/or television stations get the largest audience share.

detention area—depressed areas designed, like the retention area, to handle rain water on a developed site.

destination point—an area that, because of development, provides the population with a reason to stop there. Specifically, an area with more than one place to shop, eat, or be entertained.

developer—the person who builds on a piece of property or takes a piece of property and properly splits it into lots for commercial or residential use.

earnest money—money given as a pledge in a binding contract.

easement—the right to go onto someone else's property. The purpose is usually specified, i.e., a utility easement would indicate that a utility such as a gas line could be extended across your property.

egress—an approved exit from your property.

elevation—the altitude above sea level of a piece of property.

endangered tree—a tree or trees that are in danger of becoming extinct or are dying off.

engineer—a person trained, preferably licensed or certified, in a technical field such as a civil engineer who plans and builds roads and bridges or a utility engineer who plans and builds utility systems.

escrow deposit—money placed in the care of a third party until certain conditions are met.

excavation—used in this text to mean the removal or cleaning out, as in excavation of trees or rock formations.

farm belt—a zoning classification identifying an area set apart for farming use. This area or zoning has specific usage and development restrictions and usually has a much lower tax base. Similar to greenbelt zoning.

feasibility study—an outline of the costs of your project matched against potential sales or incomes.

fill—dirt used to fill holes or low areas, or to grade property.

fixed rate mortgage—a mortgage that carries the same interest rate throughout the life of the loan.

flood plain—an area identified as having had a flood or is likely to have a flood if specific conditions occur.

franchise—exclusive rights to an idea, business, marketing program, or territory, etc.

franchisee—a person or business that has been granted a franchise.

franchisor—a company that grants a franchise.

garbage dump—an area that was previously used as a garbage dump which was buried, and later filled up. These dumps are not always detectable and over time when decay begins, the ground can settle, causing cave-ins and other serious damage to anything built on it.

generators—in this text, this term is intended to mean something that attracts business, i.e., a shopping center is a customer "generator" for a bank or restaurant because it draws people to the area or destination point.

going home side—the side of the road most used as the main route from work to home.

grading—to move dirt around to achieve a desired ground level or contour.

grandfathered-in zoning—zoning in place before a zoning change. Usually the property is being used as previously zoned.

greenbelt—specifies an area zoned for farming or agriculture. This zoning usually has a very low tax base.

growth area—an area that is rapidly being developed. This development could be residential, commercial, or industrial.

historical structure—the internal revenue service allows tax advantages for nonresidential commercial structures built before 1936 and structures certified by the national register of historic places. Also, structures that are recognized, and especially certified, can have heavy restrictions on remodeling or any other changes.

horticulturist—a person who specializes in the art or science of plants and flowers.

impact fees—fees charged by the utility department to use the utility system. Impact fees can also include road impact fees for the use of the road.

improved property—property that has something built on it.

improvement—A building or property that has been improved by putting in roads and utility lines, etc.

independent utility service—a utility system independently owned and operated rather than owned and operated by the city or county.

infrastructure—the roads and utility lines of a project. Can also include sidewalks, special grading, landscaping, and other development requirements of the area.

ingress—an approved entrance to your property.

leasing agent—someone who handles leasing for apartments, shopping centers, etc.

lender—bank or anyone who loans money.

liability insurance—insurance that covers you if you are responsible for damage to someone's property or body. For instance, if you own ten acres and someone walks across it and falls and breaks a leg, you are liable for their injury.

lift station—a system designed to boost the gravitation of the sanitary sewer system.

M.A.I.—member appraisal institute, nationally recognized. A designation awarded by the American Institute of Real Estate Appraisers. Most lenders do not accept appraisals done by anyone but an M.A.I.

mangrove—vegetation (small trees) that grow in swamps along the banks of rivers. Mangroves are thought to protect the shoreline, and in some areas, it is against the law to remove them.

man-made barrier—a barrier made by man such as a bridge or interstate highway.

market area—the geographical area that your customers are in.

marketable title—a title that is clear, well thought out, and that can be sold.

marketing agent—a person who handles the advertising, and sometimes sale, of your project.

median—a separation of a street or highway, sometimes concrete or landscaping.

median cut—an interruption in the median for turning or entering a place of business or accessing another road.

mineral rights—an owner can retain mineral rights to a piece of property he sells, or he can sell these mineral rights to another party. This can mean that the owner of the mineral rights can come on the property, look for minerals and remove them.

mingle—a word meaning to indicate the unrelated persons who team up to lease or purchase an apartment or house.

moratorium—a period in which something stops. In this text it is used to mean that no building is allowed, or no utilities are supplied, or some other service, needed for site development.

muck—very wet soil that will not hold a building or any type of development.

natural barrier—a barrier that is not man-made, such as a mountain or river.

option—an agreement to purchase property at a certain price at a certain time.

outparcel—a term used by shopping center developers for the sites available at the shopping center site as part of their project.

overdeveloped—(Also called overbuilt.) When there is an over supply of projects, i.e., more apartments than needed, or more single family houses, than there are people to buy or rent them.

planned unit development (PUD)—when a developer buys a large piece of property (50+ acres), and plans mixed-uses for it, such as 30 acres for single family homes, 10 acres for apartments, and 10 acres for commercial use, the permitting and zoning approvals recognize this property as a planned unit development. Because the developer puts in roads and utilities etc., the zoning and planning is handled differently.

plat—a parcel of land surveyed and divided into lots.

plot—a piece of land

reciprocal cross parking easement—an agreement that not only allows you to park on someone else's property, but also allows them to park on your property.

retention area—when a piece of property is developed, and a building and parking lot added, it can no longer hold or absorb rain water. To prevent water from draining to neighboring properties, in some areas of the country, you must put in a retention pond to catch and hold the same amount of water the property held before development.

road impact fees—in some areas there is a fee for the roads when you develop property. The idea behind the road impact fee is that whatever you are putting on the property will increase wear and tear on the road.

SMSA—(standard metropolitan statistical area.) The counties, cities and towns surrounding a large city which make up the population.

septic tank—an underground holding tank in which sewage is reduced to liquid by bacterial action. The liquid then drains out of the tank into drain areas or drain fields.

setback—a term used by the building and zoning department to determine how many feet from the property lines a building or house must be. Setbacks are a minimum and usually can be increased. In most cases, however, it cannot be decreased.

sewer plant—a plant set up to purify the sewage before it is released to the drain field.

sign ordinance—the rules and restrictions pertaining to the size, color, placement, height, and type of sign.

sinkhole—a surface area which has sunk, leaving a hole in the ground that usually fills with water. In some instances, a soil analysis can determine conditions conducive to a sinkhole.

site selection—the act of finding a piece of property for your needs.

site work—the work done to clear a site of trees and vegetation, prepare the soil for a particular purpose, including installing utility lines, etc.

soil test—a test done by a soil company to determine the type of soil of a piece of property. The depth of the test depends on your needs and the area.

storm sewer—an underground system where rain drains and is carried away to prevent flooding.

subdivision—a parcel of land that has been developed into lots for houses. Usually restrictions other than zoning requirements are included, i.e., some subdivisions do not permit a clothesline in the yard.

surface drainage—this term most often refers to rain water. A site should have a system to handle the drainage of the property. Usually this is done by retention ponds or a city or county system with an underground storm sewer.

survey—a system and process set up to measure the size, location, and boundaries of a particular property. It can include buildings, mountains, lakes, and any other characteristic of the property. Findings are documented with all information to-scale.

surveyor—a person trained and certified in surveying.

swamp—an area of wet, marshy property, usually very difficult to develop. Some swamps are protected and cannot be disturbed.

tax assessor—a person who appraises your property for real estate tax purposes.

tax assessor's office—the office where all the appraisals for real estate tax purposes are kept.

tenant—a person or business who rents property from someone.

timing—timing by my definition is when you have all your ducks in a row at just the time someone needs a big pot of duck soup. Timing is planning your project (supply) to coincide with the need for it (demand).

title—proof of ownership to a piece of property.

title commitment—usually at a closing, the title company gives you a title commitment which says this title has been researched and is good; we are making a commitment to you that we will write title insurance for this property and you should have your policy within the next 10 days.

title insurance—an insurance policy for a title which covers losses caused by defects in the title that show up after closing.

title search—the process the title insurance company goes through to check all transactions on the title to the property to determine if there are any problems with the title.

topography—the surface features, elevations, and other physical features of the property, including a written verification of these features.

toxic waste or hazardous material dump—areas previously used to dump toxic waste or hazardous material and then plowed under or covered with top soil. These areas, like the old garbage dumps, can give off health threatening fumes, or cave-in.

traffic bound—this term is used to describe traffic that backs up and blocks the entrance on a property or establishment.

traffic count—the number of cars per day traveling a particular road or street.

traffic impact fees—fees charged for the use of an existing road. The reasoning behind this is that when a piece of property is developed it increases the traffic, and wear and tear for that road, and the developer is charged a portion of the cost of maintenance.

tree ordinance—a tree ordinance indicates that a city and/or county has a specific plan for trees, i.e., some cities and counties are on a ''save-the-trees'' campaign and you cannot cut down a tree without a permit that complies with the tree ordinance; or, if you must remove a tree because it is right where you want your building, you can sometimes do so if you replace it with a tree elsewhere on your project.

unimproved property—property that has not been built on, cleared, or filled. It has been, for the most part, left in the original state.

user—in real estate, there can be two types of buyers: one buyer purchases the property just to resell at a later date, and the other buyer purchases the property to use it. For instance, a large manufacturing company that buys property to build a warehouse is a user. An investor who buys a piece of property, rezones it and sells it, does not use the property himself, but resells it.

user pie—a term used to describe a group of users.

utility—a utility is a service such as gas, electric, water, and sewer systems.

utility capacity—a utility plant or company can only handle so much service, once it has reached its peak, the capacity is gone. Utility capacity refers to how much service is left on a utility service or plant.

utility engineer—an engineer trained and certified in the planning, building, and construction of a utility system.

variance—a system to change or alter a zoning restriction due to the nature of the improvement.

visibility—whether or not you, your sign, or your business can be seen by the customer.

water district—an agency set up to control development of property in such a way as to prevent damage to the natural drainage of property and the natural flow of waterways throughout the state.

well—if you do not have access to a city or county water system, in some areas, with a permit, you can dig down to the natural water supply and with a well pump use this water supply.

wetlands—This is another name for swamp. An area of very wet soil, practically impossible to develop.

zoning—government regulation of land use to control the development of land. Zoning is intended to control and monitor development for the good of all.

Index